Food Safety Les
for Cannabis-Infused Edibles

Food Safety Lessons for Cannabis-Infused Edibles

Kathy Knutson, Ph.D.

Principal
Kathy Knutson Food Safety Consulting LLC,
Green Bay, Wisconsin, United States

ELSEVIER

ACADEMIC PRESS
An imprint of Elsevier

Academic Press is an imprint of Elsevier
125 London Wall, London EC2Y 5AS, United Kingdom
525 B Street, Suite 1650, San Diego, CA 92101, United States
50 Hampshire Street, 5th Floor, Cambridge, MA 02139, United States
The Boulevard, Langford Lane, Kidlington, Oxford OX5 1GB, United Kingdom

Library of Congress Cataloging-in-Publication Data
A catalog record for this book is available from the Library of Congress

British Library Cataloguing-in-Publication Data
A catalogue record for this book is available from the British Library

ISBN: 978-0-12-819512-3

For information on all Academic Press publications visit our website at https://www.elsevier.com/books-and-journals

Publisher: Charlotte Cockle
Acquisitions Editor: Patricia Osborn
Editorial Project Manager: Andrea Dulberger
Production Project Manager: Kiruthika Govindaraju
Cover Designer: Alan Studholme

Typeset by TNQ Technologies

Contents

CHAPTER

Introduction

1

Introduction to the author: it's personal

I work in the cannabis industry, and I don't get high

As a food scientist, I am used to writing in the third person about science, not about me. In scientific and technical writing, we are trained to not use the word "it" and to write about the methods, observations, and analysis of data. Well, this book will have plenty of observations, but they will be my observations and, in some cases, a single data point. I will ask the questions I had about the cannabis industry, the questions that you probably have too. I will describe my lessons learned in getting the answers to those questions. I speak only of which I know. I am a bucolic girl from Wisconsin, a foodie, a locavore, and a cook. Training of adult learners is my career passion.

Note

Definitions of vocabulary start in a later section of this chapter. Turn ahead for definitions if you need them sooner.

My first exposure to weed was at a dry cleaner

Nearly everyone has a personal story of their first exposure to cannabis. Until 2017, I had almost no exposure to marijuana in my 50-plus years of life. I grew up in a small town outside Madison, Wisconsin. My first job was working six hours per week at the one dry cleaner in town and closing up after the full-time ladies finished. I was 16 and alone when one day a high school teacher of mine came in to pick up his dry cleaning. I knew him well from after-school activities he supervised. It was custom to staple anything found in the pockets of the clothes to the paper ticket stapled to the plastic bag. When I retrieved his dry-cleaned sport coat from the turning rack, there on the ticket was stapled a baggie of weed. As I did with all the customers, I simply asked if the stapled item belonged to him. He was not able to get a yes or a no out among the stammering, but he paid the dry-cleaning bill and left with the sport coat and baggie of weed. At that time, I literally had no idea it was weed. Given his reaction, it is only my logical conclusion that it was weed. It could have been oregano.

Food Safety Lessons for Cannabis-Infused Edibles. https://doi.org/10.1016/B978-0-12-819512-3.00001-8

In the last week of high school, I was at a friend's house and smelled something funny. I followed the smell and found her older brother with a few friends in the living room smoking weed. He asked me if I wanted to try it, and I said no. Beside the one time, I was never exposed to weed or asked to smoke weed through my entire K–12 education and 10 years of university education. I just did not have friends who smoked. I grew up in a time when cigarettes were advertised on billboards and in magazines. Kids smoked. Adults smoked. Restaurants and bars were smoke-filled. I tried a cigarette once. That hurt my lungs. Most of my extended family did not smoke cigarettes. Again, I just did not have friends who smoked cigarettes, yet alone weed.

Fast-forward through jobs, marriages, and raising my two sons, and "someone" in my house tried to flush weed down the toilet and it floated. "Someone" left a mostly empty bag of weed on the passenger side of my husband's car, after "someone" used the car to drive with a friend to an amusement park. I was not amused. Thus began my journey of learning about cannabis use. They say education begins at home.

In 2017 I was talking with my colleague, Rick Biros of Innovative Publishing, at the Food Safety Summit. Innovative Publishing Company LLC owns Food Safety Tech and manages the Food Safety Consortium annual meeting and exhibits. I have known Rick since 2000 and my days as Outreach Manager for the National Center for Food Safety and Technology (NCFST), renamed as the Institute of Food Safety and Health. Rick was telling me that his son, Aaron Biros, had graduated from college and joined his company to write a cannabis publication. The thought was so foreign to me that I burst out loud laughing. Oops. Rick kindly explained the role of the *Cannabis Industry Journal* and how the cannabis industry was growing. That was my first lesson.

In the fall of 2017, I received a call from another NCFST colleague, Sanford Wolgel. Sanford was looking at writing a state application for a cannabis-infused edible manufacturer and wanted to know if I was available to help with the Good Manufacturing Practices and maybe other documentation such as Standard Operating Procedures. That was my second lesson. While the project with Sanford did not go to fruition, I listened to and learned everything I could from Sanford about the cannabis industry. He had worked in laboratory operations and product development, knew the chemistry of cannabis, and still is an expert in regulations.

Note

The words *product* and *products* are obviously the singular and plural forms. However, use of *product* can be plural as well. In other words, products is plural. *Product* can be either singular or plural in use. English is fun.

My role as a consultant

While most of my career has been in academia and laboratory operations, I dipped my toe into the consulting pool in the beginning of 2016, when I became employed by Jeff Kornacki of Kornacki Microbiology Solutions, Inc. In addition to my work with Jeff, I was a Lead Instructor for the Food and Drug Administration (FDA) Preventive Controls for Human Food (PCHF) rule in the training of Preventive Controls Qualified Individuals (PCQIs) and a member of the Expert Services Team at ConnectFood where I assisted in writing food safety and Hazard Analysis and Critical Control Point (HACCP) plans. In this gig economy, I was looking for more work as an independent consultant. I thought I could take my food safety expertise and transfer my skill set to cannabis-infused edibles. I will talk more about transferring your skill set to the cannabis industry in a later chapter. The next step was to call Rick Biros. Rick got me connected to Aaron Biros, and my inauguration into the cannabis industry started with the publication of a series of my articles in Aaron's online journal, the *Cannabis Industry Journal*. That was my third lesson.

Box 1.1 Preventive controls

Preventive and *preventative* are synonyms, and the FDA chose to use *preventive*. Preventive controls are procedures to control identified hazards in a food or cannabis-infused edible facility. Chapter 6, The Hazard Analysis, and Chapter 7, Preventive Controls, will detail the steps for writing a food safety plan to include a hazard analysis and preventive controls.

I was ready to learn more about the cannabis industry. I continued to talk with Sanford and Aaron about the industry and research online. The perfect opportunity came for me in 2018 to attend a major cannabis conference and exhibit. I was scheduled to deliver the PCQI course close to San Jose just ahead of the annual conference managed by the National Cannabis Industry Association (NCIA), called the Cannabis Business Summit & Expo. As an exhibitor, Aaron had an extra badge for me to get in without paying, and my flight was covered by the training I delivered, so I attended. At the time, I was talking up the application of the FDA requirement for a written food safety plan and implementation of preventive controls because that is what I know. Most of the cannabis industry was not there yet, so I did not gain much traction, but I learned a lot about the entire industry, not just the edibles industry. That was my fourth lesson, and my knowledge and experience kept growing from there through networking and eventually projects in the cannabis industry.

Purchasing cannabidiol

Learning about cannabis is personal. For over 50 years, I have lived every day with an autoimmune disorder, similar to rheumatoid arthritis. I have inflammation, pain, and fatigue. With the cannabidiol (CBD) market penetrating everywhere in the United States, even folksy Green Bay, Wisconsin, where I live, and reading and hearing about the medical benefits of CBD, I purchased a CBD tincture in the fall of 2018. A tincture is an oil that is portioned out with a dropper and consumed orally. If you hold the tincture under your tongue as long as possible, the effect will be quicker as the CBD absorbs into blood vessels. Otherwise, swallowing the tincture starts the path of the digestive tract to the circulatory system to the liver for metabolism there. The effect then takes an hour plus or minus. Like alcohol consumption, an empty or full stomach makes a difference. I experimented with time of day and number of drops. Too late in the day made me stay up too late and not get enough sleep. Too much of the tincture put me into a deep afternoon nap. CBD is not a cure for me but is part of my assault on my disorder.

I had purchased the CBD at a nutrition store and later noticed a whole store in Green Bay dedicated to the sale of CBD and CBD products. It amazes me that a whole business can subsist on the sale of CBD. CBD is extremely expensive at \$50–\$75 for about 0.4 oz or 45 mL for the tinctures I have purchased. When talking with the store owner and manager, I was told the clear majority of sales goes not to young people, but to people like me in their 50s or older for the relief of pain. My husband reminded me also of the "discretionary income" that I have, but a young person may not have, which may be a reason why middle-aged to elderly individuals purchase CBD products.

I am now often asked what CBD product or brand a person should buy. Right there is a reason the United States needs FDA oversight of marijuana and cannabis products because I should not and do not dish out medical advice. CBD products are everywhere in retail stores such as supermarkets, nutrition stores, pharmacies, doctors' offices, and gas stations. The availability of CBD products is similar to the availability of over-the-counter (OTC) pain medications. Small specialty brands of OTC drugs compete with large name brands and generic brands. Some consumers have brand loyalty and only purchase from the large brand names. Our language uses the brand name as a synonym for the medication when we ask for Tylenol, Excedrin, or Motrin. Other consumers go for the lower cost generic OTC medication, feeling confident that the same drug is offered in both the name brand and generic medication.

I have brand loyalty to the Alive brand of vitamins I purchase from Nature's Way. The company is a national company headquartered in my home of Green Bay. I have researched the product and know people who work at Nature's Way, so I am an informed and confident customer. When shopping, I do not compare vitamin products or cost; I just buy the Alive vitamins. A CBD product can be purchased by brand or by cost. A quality CBD product should be expensive. If the CBD product is cheap, that's a red flag of the quality. When comparing brands, there are so many new companies in the marketplace that the reputation of a new company versus an old company is meaningless to me. I admit there may be a slight advantage of buying from an older company that has gone through the growing pains of the industry, but even an established company is a new company because the industry is so young. As you read on, you will see that there are many cannabinoids in addition to CBD and each may have a role in addressing health concerns on an individual basis. I have heard wonderful success stories for both humans and animals with the use of CBD. Consumers should do as much research as possible, before purchasing a CBD product, on the company and on the benefits of CBD as a remedy for their personal health issues. Seek professional resources of information independent of any one company, and be wary of any company advertising a health claim as that is absolutely illegal without expensive drug studies to support the claim.

You might think it is simple to compare the cost of CBD product brands in an apple-to-apple kind of way. My limited experience has been frustrating. Purchasing a larger volume or weight of CBD product may or may not be a cost savings. For example, in buying a bottle of CBD tincture, the variables are cost, total volume, total CBD per bottle, serving size, and CBD per serving. I can easily calculate cost per volume of two different sizes of bottles. However, the important information is the dosage of CBD. If I want a dose of 10 mg of CBD and the serving sizes and CBD per serving vary between two bottles, it gets confusing. A typical serving size is 1 mL, which is 20 drops, and the tincture is delivered with a dropper. My current CBD tincture is 25 mg per 1 mL or 20 drops. The math tells me that a dose of 10 mg is in eight drops. If I compare that tincture to a different brand, I may need only five drops of the different brand. Once I figure out the number of drops that delivers the dose of 10 mg, I calculate the cost for each tincture for the dose to choose the cheaper option. I am good at math but standing in the store and trying to figure that out is too much. This leads to my personal experience: buy a bottle and try it.

Field trips

In 2018 I started visiting dispensaries when I traveled for work into a state with legal recreational cannabis. My first trip to a dispensary was in Seattle, Washington. I consider these visits like field trips in school. As part of my job, I simply must go to dispensaries to learn more about the cannabis industry! When Illinois writes and implements legislation for legal recreational cannabis, it will be the 11th state in the United States to do so. Until then, that only gives me 10 states where recreational cannabis is legal. I have been to a dispensary in Seattle, San Diego, and Denver. As a special guest, I visited a dispensary in Florida, but I could not purchase. Florida only has medical marijuana, along with 32 other states, which requires a state-issued medical card to enter the dispensary. So you see my field trip options are limited. Living in Wisconsin, we do not even have medical marijuana, and I do not have access to medical marijuana for my pain.

December 24, 2018, will live in infamy for me as the day I first inhaled Δ^9-tetrahydrocannabinol (THC). I had already had too much Christmas wine, so I felt nothing from the THC. I have tried unsuccessfully to get high. I haven't tried very hard because I was on work trips in states where recreational cannabis was legal and that was not the time to experiment. Those stories are told in Chapter 3, Begin with the End in Mind®: The Dispensary. In Chapter 10, Operation: Packaging and Labeling of Edibles, I tell the story of my field trip in Colorado. I may have been high for 5 minutes, so in my mind that doesn't count. That's like saying I was tipsy from an alcoholic drink for 5 minutes. That doesn't count.

Box 1.2 I'm just curious, what 11 states have legal recreational cannabis?

I am trying to make this a party trick of mine—to list all 10 states with recreational cannabis. By the time you are reading this book, Illinois will be the 11th state with recreational cannabis. Using the visual aid here (Fig. 1.1), I start on the west coast and travel east: Alaska, Washington, Oregon, California, Nevada, Colorado, Illinois, Michigan, Vermont, Maine, and Massachusetts, as well as the District of Columbia. I hear Vermont is lovely, and I have never been to that state. Field trip, anyone?

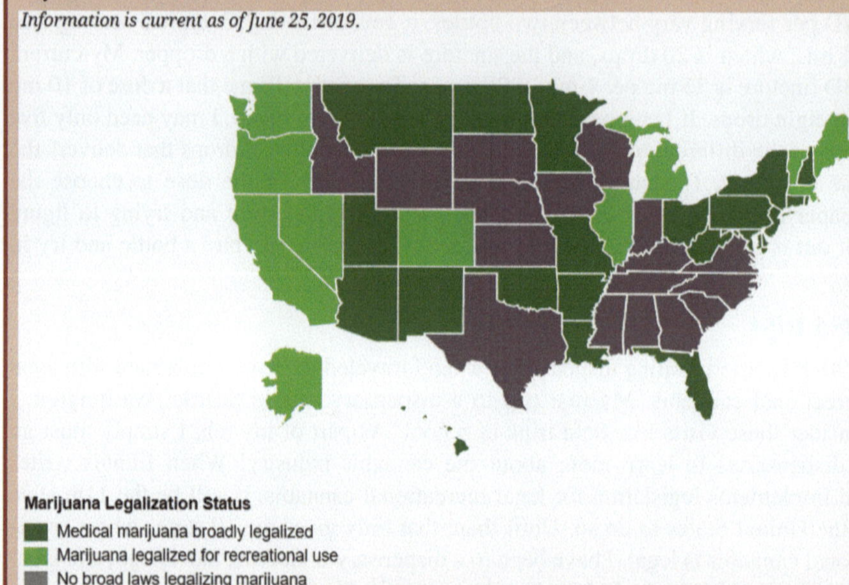

FIGURE 1.1

Illinois became the 11th state to legalize recreational use of marijuana [6]. *Used with permission from e.Republic.*

Introduction to cannabis vocabulary

To work in the cannabis industry, at some level, you need to be comfortable with chemistry and botany. At one point, I counted that I had taken 10 college chemistry courses, so I do know chemistry, but it took time to feel comfortable with the vocabulary of cannabis. Here are some basic terms everyone should know to work in the cannabis industry [1,2], and these will be used throughout the next chapters. I am thankful to Weedmaps for their comprehensive Cannabis Dictionary [3].

THC: a major cannabinoid in cannabis and causes psychoactive intoxication.

Black market: operating a business illegally. An example of a black market operation is the sale of a cannabis product to a minor. In contrast, the *gray market* is operating a business legally where laws exist, complying with most parts of the laws but not all. In the cannabis industry, an example of operating in the gray market is when cannabis-infused edibles are sold in packaging through which the edible can be seen. The state code for packaging requires opaque packaging to conceal the edible.

Bud: see Flower. *Bud* and *flower* are terms that are often used interchangeably.

CBD: one cannabinoid found in hemp and cannabis with nonintoxicating and potential medicinal or therapeutic properties. Cannabinoids that are structurally related to THC and CBD are tetrahydrocannabivarin (THCV), cannabichromene (CBC), cannabigerol (CBG), and cannabinol (CBN).

Cannabinoids: lipophilic chemical compounds that react with receptors in the brain and immune system. Over 100 different cannabinoids have been identified in the cannabis plant [4].

Cannabis: a flowering plant in the genus *Cannabis*. *Cannabis* species are *Cannabis sativa*, *Cannabis indica*, and *Cannabis ruderalis*. *C. ruderalis* may not be a separate species. Cannabis hybrids are developed with a cross of *C. indica* and *C. sativa*. In addition to the proper name of the plant, cannabis is a general term with the synonyms of marijuana and weed and its use is not italicized or capitalized.

Cannabis-infused edible: a.k.a. marijuana-infused product (MIP). Cannabis-infused edibles are cannabis products made for ingestion, including food; dietary supplements in the form of pressed pills, capsules, or teas; and tinctures.

Concentrate: the product of hemp or cannabis extraction that contains THC and CBD. Extraction may be followed by the processes of winterization, decarboxylation, or distillation.

Deschedule: to remove marijuana from the Schedule I drug list of the Drug Enforcement Administration (DEA).

Dispensary: store for the sale of cannabis products.

Edible: see cannabis-infused edible.

FDA: Food and Drug Administrtion, a federal agency in the Department of Health and Human Services. In addition to food and dietary supplements, the FDA regulates medical devices, tobacco, animal and veterinary products, cosmetics, pharmaceuticals, radiation-emitting devices, and biologics including vaccines and blood. No matter where the federal regulation of cannabis-infused edibles falls, regulations will be under the jurisdiction and enforcement of the FDA.

Flower: general term that refers to the smokable, trichome-covered part of a female cannabis plant, also known as nuggets, nuggs, nugs, or bud [5]. The cannabis flower does not have the look of a traditional flower, such as a rose, daisy, or lily, but all the botanical parts are present.

Hemp: *C. sativa* strains grown for its fiber and oil seeds. Hemp grown for hemp-derived CBD is legal with the signing of the 2018 Farm Bill.

Manifest: written record for transport of hemp, cannabis, or cannabis products.

Marijuana: slang term for cannabis. The connotation of the term *marijuana* is for the drug-like or medicinal use of cannabis.

Medical marijuana: general term for the legal sale of cannabis products to qualifying patients with a state medical card. The sale of medical marijuana is electronically tracked on a single platform separately by each state.

MIP: see cannabis-infused edible.

Recreational (use) cannabis: general term for the legal sale of cannabis products for recreational use and not medicinal use. The implication is that the cannabis product will create a high or intoxication. Consumption of recreational cannabis is analogous to the legal consumption of alcoholic beverages by adults.

Tincture: a liquid cannabis product delivered with an eye dropper for ingestion.

Weed: slang term for cannabis.

Introduction to cannabis regulation

I'm just curious, where is cannabis legal?

First, this is a follow-up question to the earlier question, *I'm just curious, what 10 states have legal recreational cannabis?* The answer was Alaska, Washington, Oregon, California, Nevada, Colorado, Michigan, Vermont, Maine, Massachusetts, the District of Columbia, and Illinois is writing and then implementing their legislation. This is a hot topic book, and the answer is changing almost monthly not only at the state level but also at the international level.

When I entered the cannabis space, this was one of my first questions and the multiple answers were overwhelming. In all my networking and in all my discussions to wrap my head around the answer, I found no one able to directly answer the question. With the rapidly changing environment of regulation, there are multiple layers of answers. Fig. 1.1 shows the US map of legalized marijuana by state. States with legal recreational use also have legal medical marijuana [6]. Here are some details for the interpretation of legalized marijuana.

1. There is a difference between legalization of medical marijuana and legalization of recreational use cannabis. Before the November 2019 ballots, 33 states had legal medical marijuana.
2. For recreational use, there is a difference between cannabis-infused edibles and cannabis for smoke, vape, dab, or any other nonedible use, including topical use such as salves or lotions. Although Canada legalized cannabis in 2018, edibles are illegal until Canada implements its regulation for edibles in October 2019.
3. Every state writes its cannabis regulations independently, so the regulation of every state is different for requirements for licensing, enforcement, sampling and testing programs, taxation, responsible state departments, and on and on. The variables are practically endless.
4. From state to state, there is a difference in
   a. the number of cannabis plants an individual may possess,
   b. the number of grams sold to an individual,
   c. the number of grams carried on an individual.

A brief history of cannabis regulations

Marihuana Tax Act of 1937

Marihuana is the correct spelling in the Act; Michigan today still uses the spelling of Marihuana in its regulations. From early colonial times, hemp was grown for its fibers in making rope and clothing. The 1906 Pure Drug and Food Act required drugs to be labeled [7] when they contained cannabis. During World War II the United States Department of Agriculture (USDA) had a "Hemp for Victory" campaign to encourage its growth for materials needed for war. However, in 1937, Congress had passed the Marihuana Tax Act essentially making cannabis for hemp and cannabis for drugs or recreation illegal, unless a tax was paid by the farmer, physician, or pharmacist.

Controlled Substances Act of 1970

The major reason we are where we are in the United States with marijuana regulation is the Controlled Substances Act. In 1970 with the Controlled Substances Act and in 1973 with the creation of the DEA, marijuana was placed in the most severe drug category of Schedule I along with heroin, lysergic acid diethylamide (LSD), 3,4-methylenedioxymethamphetamine (ecstasy), methaqualone, and peyote with the definition of being drugs with no currently accepted medical use and a high potential for abuse [8]. When President Nixon declared his war on drugs, marijuana was included with heroin and LSD. Once marijuana was on the list, it was never removed.

Decriminalization

Decriminalization of cannabis is when possession of small amounts of weed does not lead to jail or prison time but may result in a fine. Possession for personal use is not a criminal act if the state laws are followed [9]. In the 1970s, states started to decriminalize cannabis, and in 1996, California was the first state to legalize medical marijuana [10].

Marijuana Freedom and Opportunity Act of 2018 [11]

As reported by the *Cannabis Industry Journal* [12], the Marijuana Freedom and Opportunity Act of 2018 was introduced by Senator Schumer on June 28, 2018:

> *This bill removes marijuana from the list of scheduled substances under the Controlled Substance Act and eliminates criminal penalties for an individual who imports, exports, manufactures, distributes, or possesses with intent to distribute marijuana. Additionally, the bill does the following:*
>
> - *establishes a trust fund to assist women-owned and minority-owned marijuana businesses,*
> - *requires federal research on the impacts of marijuana use on highway safety and public health,*
> - *authorizes federal restrictions on the marketing of marijuana-related products, and*
> - *authorizes grants for state and local governments to expunge or seal convictions for marijuana possession.*

The act was referred to the Committee on the Judiciary.

Agriculture Improvement Act of 2018 [13]

It was the signing of the Agriculture Improvement Act of 2018, better known as the Farm Bill, in December that legalized the growing of hemp. The growing of hemp under the Farm Bill is heavily controlled. The cannabis strain for hemp must have no more than 0.3% THC. With the Farm Bill signed, the DEA moved products derived from hemp from the Schedule I list of drugs to the Schedule V list, the least severe drug category [14]. Marijuana remains on the Schedule I list. CBD is now a Schedule V drug with the following caveat from the DEA for approved products [8]:

> *A drug product in finished dosage formulation that has been approved by the U.S. Food and Drug Administration that contains Cannabidiol derived from cannabis and no more than 0.1 percent (w/w) residual tetrahydrocannabinols.*

The Farm Bill cut-off is 0.3% THC in hemp, and the DEA cut-off for a CBD drug is 0.1% THC. There is only one FDA-approved drug with CBD, which is Epidiolex [15]. Many believe the Farm Bill and DEA action have legalized the sale of CBD. Not really, and yet hemp-derived CBD products are everywhere being sold in retail stores. The FDA is currently operating under enforcement discretion, meaning the FDA will not stop the sale of CBD products [16]. The FDA was feeling the pressure and held a public hearing on May 31, 2019, to listen to public comments on the regulation of cannabis products. There is an FDA working group charged with addressing the public comments [17].

Descheduling is the future

The future is descheduling of marijuana from the DEA Schedule I list at the federal level. To that point, the American Bar Association passed resolution 104 stating [18],

> *RESOLVED, That the American Bar Association urges Congress to enact legislation to exempt from the Controlled Substances Act any production, distribution, possession, or use of marijuana carried out in compliance with state laws; FURTHER RESOLVED, That the American Bar Association urges Congress to enact legislation to remove marijuana from Schedule I of the Controlled Substances Act, 21 U.S.C. §§ 801 et seq.; and FURTHER RESOLVED, That the American Bar Association urges Congress to enact legislation to encourage scientific research into the efficacy, dose, routes of administration, or side effects of commonly used and commercially available cannabis products in the United States.*

The NCIA and other professional associations have strong lobbying efforts to push the government to move on descheduling. NCIA's Andrew Kline, the NCIA Policy Council Director, detailed the meaning of descheduling of cannabis [19]:

- *Removing cannabis from the authority of the DEA*
- *Removing cannabis from a list of illegal drugs that have no medicinal benefits like heroin, LSD, and meth*
- *Legalizing cannabis at the federal level so that there is no conflict with state laws*
- *Legalizing cannabis at the federal level so that banks no longer [have] a risk of federal money laundering charges by doing business with the industry*
- *Legalizing cannabis at the federal level so that the federal tax code permits businesses to take small business deductions*
- *And providing FDA and the Department of Treasury with regulatory authority like they do with alcohol, tobacco, prescriptions drugs, dietary supplements, and foods*

The National Cannabis Industry Association's four lanes of regulation proposal

On October 1, 2019, the NCIA Policy Council published *Adapting a Regulatory Framework for the Emerging Cannabis Industry* [20]. The white paper proposes four lanes for the regulation of marijuana and hemp products:

- Lane #1: Pharmaceutical drugs
- Lane #2: Ingested, inhaled, and topically applied THC products (>0.3%)
- Lane #3: Ingested and inhaled cannabinoid products with low/no THC (<0.3%)
- Lane #4: Topically applied cannabinoid products with low/no THC (<0.3%)

Epidiolex is the only FDA-approved pharmaceutical and belongs in Lane #1. Lane #2 THC products are only available at dispensaries in states with legal medical marijuana or recreational cannabis. Lanes #3 and #4 contain the CBD edibles and topicals, respectively, that are commonly found for sale at retail stores. The NCIA members will use the white paper for communication with federal regulators on behalf of the cannabis industry.

November 2019: Marijuana Opportunity Reinvestment and Expungement Act

A positive step in the legalization of marijuana slipped by almost unnoticed in November 2019 due to the impeachment hearings of President Trump dominating the news. The House Judiciary Committee passed the Marijuana Opportunity Reinvestment and Expungement (MORE) Act with a vote of 24 to 10 [21]. The MORE Act would remove marijuana from the federal Controlled Substances Act and require the expungement of past federal cannabis convictions among other steps. Although the MORE Act has a good chance of passing when it goes before the full House, the 2020 Senate is not likely to pass it. Democratic presidential candidate, Senator Kamala Harris has already introduced the senate companion bill [22]. All Democratic presidential candidates, except former Vice President Joe Biden, are in favor of legalization of marijuana. In a campaign speech, Biden called for more research on marijuana to rule out its role as a gateway drug, and following the November Democratic debate in which Biden was challenged by Senator Cory Booker, the campaign released Biden's stance on marijuana laws to include (1) no one should be in jail for marijuana use, (2) decriminalize recreational marijuana use and automatically expunge prior convictions, (3) legalize medical marijuana, (4) allow states to legalize recreational marijuana, and (5) reschedule marijuana so researchers can study its health impacts [23]. There are two more 2019 bills to follow: the Strengthening the Tenth Amendment Through Entrusting States (STATES) Act [24] would recognize the state laws that have legalized marijuana and the Secure and Fair Enforcement (SAFE) Banking Act [25] that opens doors for cannabis companies to bank under federal laws.

The Food and Drug Administration regulation

Historically the cannabis industry has operated with a healthy dose of avoiding government regulation and mistrust of the government. Until now, I have never felt that way and now I am starting to understand. I think I am getting a little paranoid about FDA and cannabis regulation. Part of the problem in the United States is that a cannabis-infused edible is neither a food by definition nor a dietary supplement.

What is food?

From the US Code-Title 21 [26]:

> *(f) The term "food" means (1) articles used for food or drink for man or other animals, (2) chewing gum, and (3) articles used for components of any such article.*

The last part, (3) components, means ingredients are food too. Of all the food in the world, chewing gum gets a shout out as food in the federal definition.

What is a dietary supplement?

From the FDA New Dietary Ingredients in Dietary Supplements - Background for Industry [27]:

> *The term "dietary supplement" means a product (other than tobacco) intended to supplement the diet that contains one or more dietary ingredients. A dietary supplement is limited to products that are intended for ingestion in tablet, capsule, powder, softgel, gelcap, liquid, or other form, that are not represented as conventional food or as the sole item of a meal or of the diet, and that are labeled as dietary supplements.*

Everything about this definition of a dietary supplement tells me that a cannabis-infused edible is a dietary supplement. What scares me about cannabis-infused edible regulation is that the cannabis ingredient must be approved as a new dietary ingredient [27,28]:

> *The term "new dietary ingredient" means a dietary ingredient that was not marketed in the United States in a dietary supplement before October 15, 1994.*
>
> *If you want to market a "new dietary ingredient" in a dietary supplement, you must be sure that the substance is considered to be a "dietary ingredient". (See section 201(ff)(1) of the FD&C Act (21 U.S.C 321(ff)(1)). A dietary ingredient is a vitamin; a mineral; an herb or other botanical; an amino acid; a dietary substance for use by man to supplement the diet by increasing total dietary intake; or a concentrate, metabolite, constituent, extract, or combination of any of the above dietary ingredients.*

If cannabis concentrates are dietary ingredients, there must be FDA approval. The field of food safety will overlap with dietary supplements in that dietary ingredient manufacturers are required to implement a food safety plan and have a PCQI. Dietary supplement manufacturers are exempt from the food safety plan and PCQI requirement. I disagree with that policy, and I hope FDA will take this opportunity of cannabis regulation to require dietary supplement manufacturers to implement a comprehensive food safety plan as is required by the food industry.

Here's a question I can't answer: will cannabis concentrate be regulated as a food additive, after descheduling of marijuana?

If instead of regulating cannabis-infused edibles as dietary supplements, FDA regulates edibles as food, then the cannabis concentrate may be defined as a food additive.

What is a food additive?

From FDA's Overview of Food Ingredients, Additives & Colors [29], a food additive is:

> *any substance the intended use of which results or may reasonably be expected to result – directly or indirectly – in its becoming a component or otherwise affecting the characteristics of any food. This definition includes any substance used in the production, processing, treatment, packaging, transportation or storage of food. The purpose of the legal definition, however, is to impose a premarket approval requirement. Therefore, this definition excludes ingredients whose use is generally recognized as safe…*

The key term in the quote is *premarket approval requirement*, which of course follows an FDA process. As a food additive, FDA may set a defect action level for CBD and THC in food. Any cannabis product over the defect action level would be adulterated. Finally, there is one more option for the FDA and that is to label cannabis-infused edibles as functional foods.

What is a functional food?

In the difference between "Food" and "Functional Food" and the Substantiation of Associated Claims [30], functional foods are defined as:

> *those foods or ingredients that supply a benefit beyond that of the traditional (i.e., mainstream) nutritive value.*
>
> *Therefore, application of the term "functional" beyond that of traditional nutritive value is an expressed (not implied) claim. That is, the seller is offering a product that will affect the structure or function of the body, otherwise it would not be "functional" and no one would purchase the product. It follows then, that the seller must substantiate his claim, proving the functionality through clinical trials – the type and rigor of which has increasingly become the subject of controversy and confusion between the two controlling Federal agencies, the Food and Drug Administration (FDA) and Federal Trade Commission (FTC).*

I can see where cannabis-infused edibles absolutely meet the definition of a functional food. Many health claims for CBD and cannabis products are illegally made due to lack of data to substantiate the claim. Clinical trials are the route needed to collect the kind of data required by the FDA for approval of a claim for a functional ingredient or food.

A tale of homonyms

In the summer of 2018, I attended the fifth annual conference of the NCIA called Cannabis Business Summit & Expo (#CannaBizSummit). I was a novice in the cannabis industry and confused about what I was hearing about flour. I thought it was cool that since I was interested in edibles, I was hearing so much discussion of flour, obviously for use in baking operations. By the end of the conference, I figured out that flour was really "flower." That is cannabis flower, not flour from cannabis. Flour is not made from cannabis, yet.

One of my colleagues attended the FDA public hearing on cannabis on May 31, 2019, also as a novice to the cannabis industry. I had listened to the first half, so I was familiar with some of the talks. On a follow-up call with her, she was adorable when she asked what the speakers were talking about when they kept referring to grass. She wondered if grass was a slang term like weed. What she had heard as grass, I heard as GRAS (generally recognized as safe). The FDA maintains a list of safe food ingredients that are GRAS.

Opinion on cannabis regulation

Cannabis-infused edibles will be regulated as a food, as a dietary supplement, or as some combination of both. I pray for the FDA to regulate cannabis-infused edibles with existing regulation. Knowing what I know now about cannabis regulations, and I admit that is limited, this is how I think of cannabis regulations, in general: let's not reinvent the wheel here. Like myself, there are many experts in their fields crossing over into the cannabis industry. Like you, they are regulatory affairs professionals, laboratory managers, chemists, microbiologists, inspectors, quality assurance managers, engineers, logistics managers, sales professionals, and marketing professionals. They are coming from the food industry, pharmaceuticals, agriculture, horticulture, dietary supplements, pet food, food service, and more. After descheduling of marijuana can we marry existing federal regulation to cannabis licensure in the cannabis industry? Let's take a look at existing agencies and their responsibilities for growing operations and harvesting, processing and manufacturing, dispensaries, transportation and laboratories.

License for growing operations and harvesting

Environmental Protection Agency (EPA)

- Registration of pesticides and sanitizers
- Clean water intake
- Wastewater discharge
- Clean air intake and discharge

United States Department of Agriculture (USDA)

- Application of pesticides
- Agricultural water quality and use
- Soil amendments

Occupational Safety and Health Administration (OSHA)

- Worker safety

Of course, OSHA rules for worker safety will be applied at every step in the seed to sale continuum.

License for drying, extraction, processing, and manufacturing operations

Food and Drug Administration (FDA)

- Thermally Processed Low-acid Foods Packaged in Hermetically Sealed Containers (21 CFR 113)
- Current Good Manufacturing Practice in Manufacturing, Packaging, Labeling, or Holding Operations for Dietary Supplements (21 CFR 111)
- Juice HACCP rule (21 CFR 120)
- Food Allergy Labeling and Consumer Protection Act
- The seven foundational rules of the Food Safety Modernization Act:
  1. Accreditation of Third-Party Certification Bodies to Conduct Food Safety Audits and to Issue Certifications
  2. Current Good Manufacturing Practice, Hazard Analysis, and Risk-Based Preventive Controls for Human Food
  3. Current Good Manufacturing Practice, Hazard Analysis, and Risk-Based Preventive Controls for Food for Animals
  4. Foreign Supplier Verification Programs for Importers of Food for Humans and Animals
  5. Mitigation Strategies to Protect Food Against Intentional Adulteration
  6. Sanitary Transportation of Human and Animal Food
  7. Standards for the Growing, Harvesting, Packing, and Holding of Produce for Human Consumption

The last rule for produce does not apply to cannabis. Produce is harvested and consumed raw, e.g., lettuces, fruits, and sprouts. Because cannabis is dried after harvesting, the produce rule does not apply. The FDA produce rule has requirements for agricultural water and soil amendments, which can apply in the above mentioned category of License for Growing Operations and Harvesting.

No matter where cannabis-infused edible regulation falls, it will fall under the jurisdiction and enforcement of the FDA. Other cannabis products will likely fall under FDA as well, to include products such as salve, lotion, lip balm, pet supplements, rolled joints, dried flower, dab, and shatter. New products are coming to market every day. Once marijuana is descheduled, we can expect FDA health campaigns to include cannabis products, similar to the current campaign to inform the public of the hazards of vaping and to reduce teenage vaping [31].

License for dispensaries

Food and Drug Administration(FDA)

- Food Code

License for transportation

Food and Drug Administration(FDA)

- Sanitary Transportation of Human and Animal Food

U.S. Department of Transportation (DOT)

The sale of cannabis-infused edibles not only will be at brick-and-mortar dispensaries but also will be online. In the same way I can order my prescription drugs, OTC drugs, and dietary supplements for delivery, I will be able to have my edibles delivered. There are already cannabis couriers that deliver products to customers in some states. If a patient has a medical card for medical marijuana and is limited in mobility, it is a great service to have the medical marijuana delivered to the patient.

In the food industry, we are used to the chain of custody for shipping, receiving, and handling of our ingredients and products sent for testing. The cannabis industry is a whole different game. A manifest is created for the movement of cannabis ingredients, product, or samples. Before shipping cannabis samples to a cannabis laboratory, the state code must be followed for who collects the samples, for the sampling procedure, for documentation, and then for the transport of samples. If you have ever lived through a Class I recall with a food company, you know the detail in amount of product shipped and paperwork for disposal of contaminated product. It's a big deal to account for all contaminated product, the amount still in the company's control, and the amount disposed. Take that concept and multiply by a factor of 10 for the transport of cannabis. First, every cannabis plant in a growing operation is tagged with a bar code. Starting at harvest the weight is recorded for the plant, trim, flower, shake, and every step along the way. You can't just move a plant or a product. If a cannabis company has two buildings on the property and one is used for growing and the other is used for packaging, there is no simple carrying of dried cannabis across the parking lot. The product would be weighed, labeled, secured, and moved with a manifest. The same is true for transport in a vehicle. It is best to have two drivers for accountability. Right there the cannabis industry doubles the payroll for drivers, compared with the food industry. Some companies have armed drivers; some companies have armed security for deliveries. Imagine receiving an ingredient at your facility that required the presence of an armed security guard for the safety of receiving. My cousin has told me stories of delivering goods to convenience and grocery stores in Chicago. One of the sought-after commodities to steal was cigarettes that would then be sold on the black market. He fended off would-be thieves as part of his daily routine by getting an "all clear" from the store manager and locking the truck doors when he went inside the store. Once marijuana is descheduled, cannabis products will move across state lines by trailer, plane, railcar, or other commercial vehicles. Current DOT regulations will serve as the foundation.

License for cannabis laboratories

The food industry best practice for laboratories is ISO 17025 accreditation, without a federal requirement. Many state food and public health laboratories are ISO 17025 accredited, and those that are not are in the process of earning accreditation. Food laboratories cannot test cannabis-infused edibles. Expect a cannabis laboratory to be a stand-alone laboratory, even if the same company's cannabis laboratory and food laboratory are both testing chocolate bars. Those working in the cannabis industry are wise to seek a cannabis laboratory with ISO 17025 accreditation as a minimum requirement, and the state code may require cannabis laboratories to be ISO 17025 accredited to earn a license.

When you find a laboratory that is ISO 17025 accredited, it is still necessary to confirm method validation for your product. When laboratory managers cross over into cannabis from other industries, they bring the expectation of and investment in method validation, which can be a tricky research project. There are no standard methods for cannabis testing, and the plant material can be difficult to hydrate and homogenize due to waxes. The Cannabis Analytical Science Program (CASP) is composed of working groups to develop cannabis standard methods within the Association of Official Analytical Chemists (AOAC) [32].

Currently in the cannabis industry, the regulation for laboratory testing is dictated in the state code, and licensed cannabis laboratories are heavily regulated. Cannabis businesses and laboratories are struggling to meet the requirements of the state code. Most cannabis businesses are not going above and beyond the state code requirements. In other words, laboratory testing is neither customer driven nor determined through a hazard analysis. In some cases, the state, local, or city code requirements don't make sense but must be followed anyway. In all cases, testing is expensive and one of the reasons I paid $70 for 0.4 oz of a CBD tincture.

Conclusion

Let's not reinvent the wheel with regulation of cannabis products and instead look to existing regulation at the federal level.

Box 1.3 SYNONYMS IN THE FOOD INDUSTRY

The cannabis industry uses the term *kitchen* to describe the location of edible manufacturing. At first, the use of *kitchen* was confusing to me. In the food industry, we have interchangeable terms that can be confusing too.

Business—the FDA defines very small and small businesses in respective Food Safety Modernization Act rules to include all locations of a company collectively.

Company—a vague term that is usually qualified; a legal connotation.

Facility—implies a building used for processing, manufacturing, packaging, or storing.

Plant—extensive use in the food industry similar to facility, e.g., pilot plant, dairy plant, plant manager. A potentially confusing term in the cannabis industry.

Operation—can be used unqualified to mean the whole facility or can be qualified, like Packaging Operation or Growing Operation, to denote a part. It implies the use of equipment at processing, manufacturing, or packaging.

The Canadians

Safe Food for Canadians

Before leaving this chapter, I want to direct your attention to the north, to Canada. Like the United States, Canada has thriving agriculture and food processing. The food safety regulation that aligns with the FDA's PCHF rule is the Canadian Food Inspection Agency's (CFIA's) Safe Food for Canadians Act. Unlike the United States, food safety is enforced by one agency in Canada, the CFIA.

Cannabis Act

Cannabis was legalized at the federal level in Canada in October of 2018 [33]. The final regulations for edible cannabis, cannabis extracts, and cannabis topicals come into effect on October 17, 2019 [34], and edibles are expected to be available for sale in Canada in December. Until the October date, it is illegal to sell cannabis-infused edibles in Canada. I appreciate the straightforward simplicity of the language in the Cannabis Act for edible cannabis [34]. Here are excerpts:

- *All edible cannabis products will need to be shelf-stable (i.e. they must not require refrigeration or freezing).*
- *Only food and food additives will be allowed to be used as ingredients in edible cannabis products*
- *Edible cannabis products will not be allowed to contain poisonous or harmful substances, nor will it be permitted to fortify edible cannabis with vitamins or mineral nutrients.*
- *The use of meat products, poultry products and fish as ingredients in cannabis products will be generally prohibited.*
- *… an exception to this prohibition will be provided for dried meats, poultry or fish, provided … that they have a water activity of 0.85 or less at a temperature of 22°C (plus or minus 2°C) at the time they are obtained.*
- *… because of the increased risk of botulism associated with low-acid canned foods, the Regulations will prohibit the sale of edible cannabis products in a hermetically sealed container if any component of the edible cannabis has a pH above 4.6 and a water activity higher than 0.85 when the product is at a temperature of 22°C (plus or minus 2°C).*
- *The use of ingredients containing naturally occurring caffeine will be permitted in edible cannabis products, provided the total amount of caffeine in each immediate container does not exceed 30 milligrams. This amendment will allow for the use of ingredients that contain naturally occurring caffeine, such as chocolate, tea, or coffee. The use of caffeine as a food additive will, however, be prohibited.*
- *Furthermore, the Regulations will allow for a small concentration of ethyl alcohol in edible cannabis products (that does not exceed 0.5% w/w), given that ethyl alcohol is often present as a by-product in fermented ingredients or products (e.g. vinegars).*

Reading through the excerpts, I am struck by the fact that Canada is regulating edibles as a food. Mentioned elsewhere is the limit of 10 mg of THC per serving. At first glance the low potency of 10 mg of THC is conservative for the rule makers, and it is better to start low and rewrite regulation as needed. However, there is a concern in the medical community that the 10 mg of THC is limiting and may not offer therapeutic value to patients.

CFIA created a role like PCQI with the requirement of a quality assurance person (QAP) for each cannabis facility. The responsibilities of a PCQI are similar to the responsibilities of a QAP for the oversight of Good Production Practices (GPP) [35] in the same way the United States follows Good Manufacturing Practices.

Quality assurance person

> *19 (1)* A holder of a licence for processing must retain the services of one individual as a quality assurance person who has the training, experience and technical knowledge related to the requirements of Parts 5 and 6 that are applicable to the class of cannabis in respect of which activities are conducted under the licence *[34].*

Appendix C: Part 2 – Licensing requirements related to GPP [35] details the responsibilities of the QAP. What is most interesting is the required Minister's approval, meaning the documentation of training, experience, and technical knowledge must be submitted to the government for approval. A facility may have alternate QAPs who must also be approved by the government [36].

Licensing

Processing Licences

Minister's approval

> *20 (1) A holder of a licence for processing must obtain the Minister's approval before*
> - *(a) designating or replacing an alternate quality assurance person; and*
> - *(b) replacing the quality assurance person by an individual, other than by the alternate quality assurance person.*

Application — content

> *(2) The holder of a licence for processing must, for the purpose of obtaining the Minister's approval, submit an application that includes the following:*
> - *(a) the name and date of birth of the proposed alternate quality assurance person or the proposed quality assurance person;*
> - *(b) a description of the proposed quality assurance person's or the proposed alternate quality assurance person's qualifications in respect of the matters referred to in subsection 19(1); and*
> - *(c) a declaration, signed and dated by the responsible person referred to in section 37, indicating that all information provided in support of the application is correct and complete to the best of their knowledge.*

Additional information

> *(3) The Minister may, on receiving an application for approval, require the submission of any additional information that pertains to the information contained in the application and that is necessary for the Minister to consider the application.*

Well played, Canada. The world is watching for your success.

References

[1] National Environmental Health Association, Cannabis 101: Glossary of Related Terms. https://www.neha.org/sites/default/files/eh-topics/food-safety/Cannabis-101-Glossary-Related-Terms.pdf. [Accessed 09.11.19].

[2] Weed guide: A visual glossary of cannabis Terminology. https://keytocannabis.com/blogs/cannabis/photos-weed-glossary-common-cannabis-terminology. [Accessed 09.15.19].

[3] Learn/Cannabis Dictionary. https://weedmaps.com/learn/dictionary/. [Accessed 09.15.19].

[4] Cannabis and its Compounds. https://www.uclahealth.org/cannabis/cannabis-and-its-compounds. [Accessed 09.15.19].

[5] Flower. https://weedmaps.com/learn/dictionary/flower/. [Accessed 09.15.19].

[6] State Marijuana Laws in 2019 Map. https://www.governing.com/gov-data/safety-justice/state-marijuana-laws-map-medical-recreational.html. [Accessed 09.11.19].

[7] Cannabis and the constitution: A brief history of Cannabis in the US. https://www.leafly.com/news/cannabis-101/cannabis-and-the-constitution-a-brief-history-of-cannabis-in-the-u-s. [Accessed 09.11.19].

[8] Drug scheduling. https://www.dea.gov/drug-scheduling. [Accessed 09.11.19].

[9] Marijuana legalization and decriminalization overview. https://criminal.findlaw.com/criminal-charges/marijuana-legalization-and-decriminalization-overview.html. [Accessed 09.11.19].

[10] California Legalizes Medical Marijuana in 1996. https://www.greenentrepreneur.com/slideshow/292452#10. [Accessed 09.11.19].

[11] S.3174 - Marijuana Freedom and Opportunity Act. https://www.congress.gov/bill/115th-congress/senate-bill/3174. [Accessed 09.11.19].

[12] The Marijuana Freedom and Opportunity Act (Finally) Makes Its Debut. https://thecannabisindustry.org/the-marijuana-freedom-and-opportunity-act-finally-makes-its-debut/. [Accessed 09.11.19].

[13] The Farm Bill, hemp legalization and the status of CBD: An explainer. https://www.brookings.edu/blog/fixgov/2018/12/14/the-farm-bill-hemp-and-cbd-explainer/. [Accessed 09.11.19].

[14] DEA moves some CBD medicines off Schedule 1, a limited expansion of cannabis access. https://mjbizdaily.com/dea-moves-cbd-medicines-off-schedule-1-a-limited-expansion-of-cannabis-access/. [Accessed 09.11.19].

[15] FDA approves first drug comprised of an active ingredient derived from marijuana to treat rare, severe forms of epilepsy. https://www.fda.gov/news-events/press-announcements/fda-approves-first-drug-comprised-active-ingredient-derived-marijuana-treat-rare-severe-forms. [Accessed 09.15.19].

[16] FDA Chief Clarifies Enforcement Priorities For CBD Products. https://www.marijuanamoment.net/fda-chief-clarifies-enforcement-priorities-for-cbd-products/. [Accessed 09.01.19].

[17] Food and Drug Administration, Public hearing: scientific data and information about products containing cannabis or cannabis-derived compounds. May 31, 2019. https://www.fda.gov/news-events/fda-meetings-conferences-and-workshops/scientific-data-and-information-about-products-containing-cannabis-or-cannabis-derived-compounds. [Accessed 09.01.19].

[18] Resolution 104. https://www.americanbar.org/content/dam/aba/images/news/2019/08/am-hod-resolutions/104.pdf. [Accessed 09.11.19].

[19] Post-Recess predictions: looking into the crystal ball. https://thecannabisindustry.org/post-recess-predictions-looking-into-the-crystal-ball/?utm_campaign=NCIA%20Content&utm_source=hs_email&utm_medium=email&utm_content=76404771&_hsenc=p2ANqtz–vzLfnAhbJpuHGkVhmCIf3lKy3KM5QRLwYhPdqkSw_-NOB9ibBgDZwI9XTDBb1HXO0JMx1_MxqvJWtzQja7OJgszpJOA&_hsmi=76404771. [Accessed 09.11.19].

[20] Adapting a regulatory framework for the emerging Cannabis industry. https://thecannabisindustry.org/wp-content/uploads/2019/09/RegulatoryFramework-WhitePaper-Final-9.30-2.pdf. [Accessed 11.25.19].

[21] H.R.3884 – Marijuana Opportunity Reinvestment and Expungement Act of 2019. https://www.congress.gov/bill/116th-congress/house-bill/3884. [Accessed 11.27.19].

[22] S. 2227: MORE Act of 2019. https://www.govtrack.us/congress/bills/116/s2227. [Accessed 11.27.19].

[23] Debate night: Booker, Biden butt heads over marijuana. https://www.axios.com/biden-debate-marijuana-weed-gateway-drug-9d7e0dab-3cfa-4b06-a558-ef9d5bdc8a41.html. [Accessed 11.27.19].

[24] All Information (Except Text) for H.R.2093 – STATES Act. https://www.congress.gov/bill/116th-congress/house-bill/2093/all-info. [Accessed 11.27.19].

[25] H.R.1595 – Secure And Fair Enforcement Banking Act of 2019. https://www.congress.gov/bill/116th-congress/house-bill/1595?q=%7B%22search%22%3A%5B%22SAFE+Banking+Act%22%5D%7D. [Accessed 11.27.19].

[26] 21 U.S.C. § 321 – U.S. Code - Unannotated Title 21. Food and Drugs § 321. Definitions; generally. https://codes.findlaw.com/us/title-21-food-and-drugs/21-usc-sect-321.html. [Accessed 09.05.19].

[27] New dietary ingredients in dietary supplements – background for industry. https://www.fda.gov/food/new-dietary-ingredients-ndi-notification-process/new-dietary-ingredients-dietary-supplements-background-industry. [Accessed 09.05.19].

[28] How to submit notifications for a new dietary ingredient. https://www.fda.gov/food/new-dietary-ingredients-ndi-notification-process/how-submit-notifications-new-dietary-ingredient. [Accessed 09.05.19].

[29] Overview of food ingredients, additives & colors. https://www.fda.gov/food/food-ingredients-packaging/overview-food-ingredients-additives-colors#foodadd. [Accessed 09.11.19].

[30] The difference between "Food" and "Functional Food" and the substantiation of associated claims. http://burdockgroup.com/the-difference-between-food-and-functional-food-and-the-substantiation-of-associated-claims/. [Accessed 09.11.19].

[31] Lung illnesses associated with use of vaping products. https://www.fda.gov/news-events/public-health-focus/lung-illnesses-associated-use-vaping-products. [Accessed 09.15.19].

[32] AOAC Announces Working Groups on Cannabis: Leading scientists developing standards and official methods for cannabis products and hemp. https://members.aoac.org/aoac_prod_imis/AOAC_Member/ANews/2019_News/Stories/NEWS_Story052319b.aspx. [Accessed 09.15.19].

[33] The Cannabis Act: Here's what you need to know. https://www.canada.ca/en/health-canada/services/drugs-medication/cannabis/resources/cannabis-act-what-you-need-to-know.html. [Accessed 09.15.19].

[34] Regulations Amending the Cannabis Regulations (New Classes of Cannabis): SOR/2019-206. http://www.gazette.gc.ca/rp-pr/p2/2019/2019-06-26/html/sor-dors206-eng.html. [Accessed 09.15.19].

[35] Guidance document: Good production practices guide for cannabis. https://www.canada.ca/en/health-canada/services/cannabis-regulations-licensed-producers/good-production-practices-guide/guidance-document.html#ac. [Accessed 09.15.19].

[36] Cannabis Regulations (SOR/2018-144). https://laws-lois.justice.gc.ca/eng/regulations/SOR-2018-144/page-5.html?txthl=quality%20assurance%20person#s-20. [Accessed 09.15.19].

CHAPTER

Overview of the cannabis industry

2

I was not familiar with the manufacture of magic brownies with cannabis when I started my cannabis journey, but I was very familiar with making a batch of brownies at home. For three years I taught Food Science at Chicago High School for Agricultural Sciences. I had one class of sophomores and double periods for both a junior class and a senior class. I don't remember what unit we were in, but one day one of the classes was in the kitchen part of the classrooms. There were six stations with four students at each station, and each station was working independently to make a different recipe of their choice. I walked among the stations to answer questions. I made my way to the back station where two students had their backs to me, working diligently. As I rounded the first student to look at their work, I see the second student in passed her wrists into a bowl of brownie batter, kneading the batter into smoothness. I asked why she was doing that with her hands. Her simple answer was that the recipe said to "mix by hand." Of course, the recipe directions really meant mix manually with a spoon or do not use an electric mixer for brownie batter.

In my own naivety, I used to think that cannabis-infused brownies were made by adding ground cannabis flower to brownie batter and baking. Knowing a bit about cannabis chemistry and a bit about baking chemistry, I now understand that the baking process of cannabis in the batter would not be long enough at a high enough temperature to decarboxylate the majority of the tetrahydrocannabinolic acid (THCA) in the cannabis into Δ^9-tetrahydrocannabinol (THC), and alas, there would be very little "magic" in the brownie. Food scientists understand the dehydration on the surface while baking leads to the Maillard browning reaction and how the temperature will be hotter there than the interior, moist batter, but that would leave a whole lot of raw cannabis to go to waste in the batter.

Instead, the manufacture of cannabis-infused brownies, or any other bakery product, is going to either *add* cannabis extract as an ingredient or *substitute* the fat in the recipe with hemp or cannabis butter or cannabis oil. I first learned the method for household cannabis butter or oil from a very chatty Uber driver. He was on a mission to educate me during our time together on the step-by-step process. I find that most people experienced in cannabis personally or professionally love to talk about cannabis. The making of cannabis products is filled with never-ending variables and questions (Fig. 2.1).

Food Safety Lessons for Cannabis-Infused Edibles. https://doi.org/10.1016/B978-0-12-819512-3.00002-X

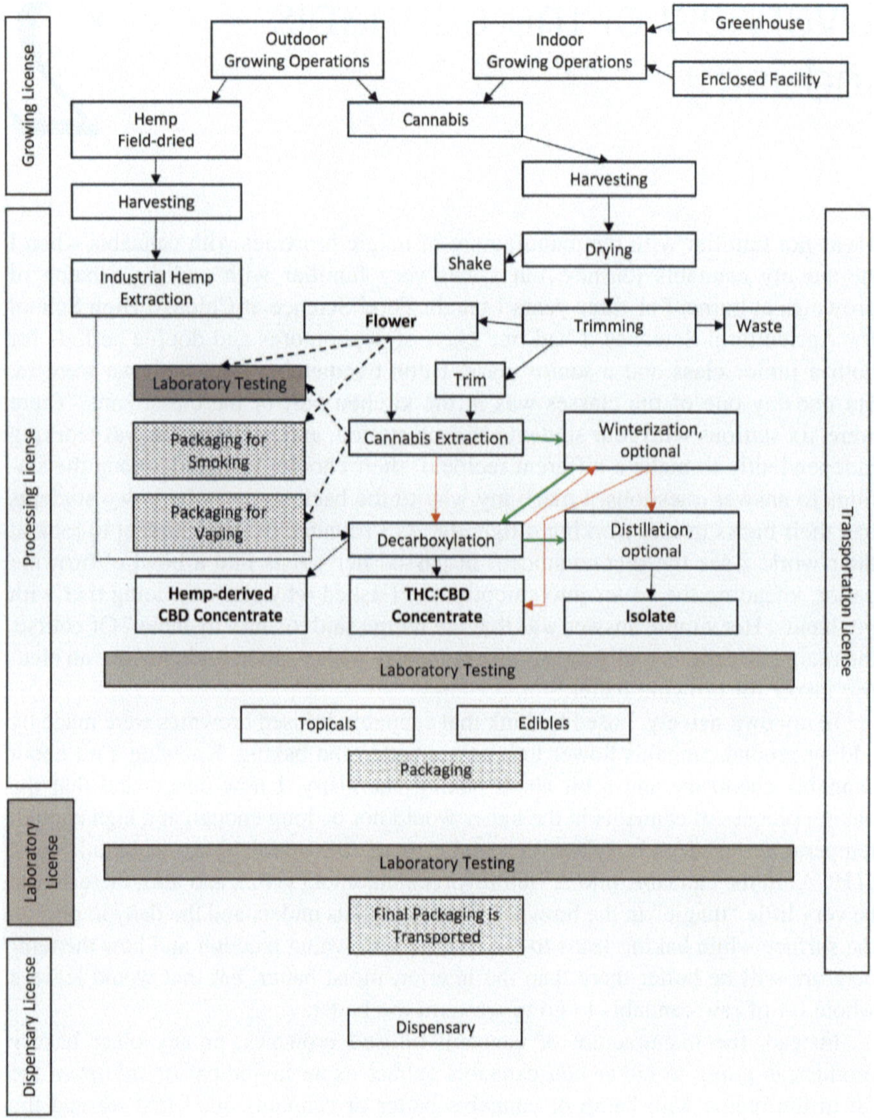

FIGURE 2.1

The growing, processing, laboratory testing, and sale of cannabis and cannabis products must follow state licensure and code. Additional laboratory testing may be required by state code than is drawn. The steps of decarboxylation, winterization, or distillation may be added for hemp-derived cannabidiol (CBD) concentrate. The steps of decarboxylation or winterization may be done in a different order of operations.

In later chapters, I discuss hazards associated with a cannabis ingredient, but first I want to paint the big picture. The cannabis industry operates seed to sale, in the same way the food industry operates farm to fork. *Farmers* grow the crop, and *suppliers* process the crop in a *de minimis* nature or function as a food facility applying significant processing steps. A *receiving facility* procures ingredients, which are mixed to manufacture a food product. If there are time and temperature parameters to meet, the process is validated. The packaged product moves to a *customer* for sale, and the product is sold to the *consumer*. In the food industry, we each know our role in farm to fork. We strive for government compliance and often second- or third-party certification. I can transfer these simple concepts from the food industry to the cannabis industry for you, but the steps in the cannabis industry are in no way simple.

Vertical integration

Vertical integration is the single ownership of all cannabis operations from growing to dispensaries, excluding independent, third-party laboratory testing. As a food scientist, when I first heard someone use the term *vertical integration*, I didn't understand. I had no experience in vertical integration. In the cannabis industry, vertical integration means that the same company owns and controls everything from the growing operation through processing and packaging and to sales at its own dispensaries. That is a lot of business knowledge to manage. Until July 2019, Colorado required vertical integration. Except for the cannabis testing laboratory, which we want to be an independent, third-party laboratory, all the steps had to be conducted and owned by one company. This is still true in Florida where medical marijuana is legal.

Imagine if a food company was responsible for growing the crop, processing and packaging the food, and selling the food to the consumer. The closest examples of vertical integration in the food industry are for some vegetables. I know companies that own the farms for growing of crops, such as sweet corn, that are picked, trucked to the factory, frozen, packaged, and sold. There are carrot farms owned by the same company that cuts, peels, and packages the raw carrots. But for perfect alignment of the food industry with vertical integration in the cannabis industry, the vegetable companies would also have to own the grocery stores that sell the frozen and fresh vegetables! As a food scientist, I am not a specialist in growing and I am not a specialist in retail sales. To impose vertical integration on the cannabis industry is unreasonable and limiting. It makes more sense to have whole businesses that specialize in one sector or step in the seed to sale continuum. Without vertical integration, businesses can specialize in what they do best, and Colorado has moved in the right direction.

Growing operations

What type of plant is *Cannabis*?

Cannabis belongs to the Family Cannabaceae with the species of *Cannabis sativa* and *Cannabis indica*; modern hybrids of the two species comprise different strains of *Cannabis* [1]. The plant is bushy with fanlike leaves being taller than it is wide. Hemp is *C. sativa* traditionally grown for its fiber and oil seeds, with the plant growing from 6 to 16 feet tall outdoors. Cannabis hybrids are grown for the desired ratio of THC:cannabidiol (CBD), and the cannabis flower of the female plant is where resin is produced on trichomes [2]. Hybrids are shorter and bushier than hemp, making them suitable for indoor and greenhouse growing. The flower is the smokable part of the plant that is dried and referred to as *bud*, *nuggets*, or *nugs*.

Cannabis is a crop

Growing of cannabis requires the same ingredients as every other crop: light, water, nutrients, time, temperature control, and humidity, with the latter being a significant factor in the case of cannabis. There are three main ways to grow cannabis: (1) outdoor in a Mediterranean-type climate, (2) within walls and ceilings with artificial light control, and (3) in a glass greenhouse with light and humidity controls. For example, for outdoor growing, there are three counties in northern California, on the coast, called the Emerald Triangle where there are good growing conditions. With artificial lighting, the best wavelength of light is used and the length of light exposure each day is programmed to get the best crop in the shortest amount of time. The glass greenhouse can use sunlight, supplement with artificial light, add humidity, and block sunlight with shades if there is too much sunlight. It doesn't matter if the structure is glass or brick-and-mortar, the growing building is called a *greenhouse*. Imagine my surprise when I visited my first greenhouse, and it was an enclosed, converted building with no windows. The cannabis industry is retrofitting commercial buildings for the growing of cannabis, which brings its own challenges. Operations with multiple facilities can afford to build big and new only after seeing success from the start-up facilities. If you are an inspector, an auditor, or have simply worked for many companies, you know the full spectrum of what constitutes a food facility from old, tiny, and packed to new and large with room to grow. Apply that concept to the cannabis industry at every step, starting at the greenhouse.

The concept of seed to sale may align with the food industry's farm to fork, but the required procedures in the cannabis industry are way ahead of the food industry. The deputy commissioner of the FDA, Frank Yiannis, promotes the use of technology for traceability such as blockchain, and the FDA will have a public meeting on October 21, 2019, on a new era of smarter food safety [3]. The cannabis industry has mastered traceability. At the greenhouse, every plant has a bar code tag around the stalk and is followed through to product. Imagine if the food industry followed that procedure for just key ingredients. Let's say cocoa processors wanted to know the source of their cocoa beans and then required growers to bar code-tag every tree to track those beans. That is the procedure in the cannabis growing operation. When the hemp or cannabis plants have fully grown, there will be harvest. In my opinion, the growing and harvesting of hemp and cannabis is no different from the growing and harvesting of any other crop and should be regulated at the federal level by the same agencies. Harvest weights are recorded for the whole plant, the discarded portion (yes, the garbage is weighed) and the desired twigs with flower going into the next step of drying. After the flower is dried, the flower is trimmed, and the portions of flower, trim, and shake are weighed and recorded. During the trimming operation, dried pieces fall off the flower as *shake* and non-flower parts are kept as *trim* (Fig. 2.1). Both trim and shake may be sold at a reduced rate compared with the more desirable flower; trim often goes into an extraction process.

I spoke with a client about using 10 plants for an internal research project. They just could not. From the start of growth, every plant is accounted for from seed or clipping to sale. If a plant does not show up at the sale step, there could be an investigation. Following growing operations, a processor has a mandatory inventory management system that tracks the product by weight. At the dispensary, each customer signs in with identification. If the customer is purchasing with a state medical card, the details of the purchase are entered in a statewide database, so the customer cannot overbuy. Imagine if adults had limits on alcohol purchase and every purchase was registered in a database. That's not exactly the right concept and purchasing of prescription drugs is close to the procedure for purchasing medical marijuana. The opioid crisis has brought to light the dangers and loopholes in the prescription drug system. The cannabis industry saw the potential for abuse through excessive purchases and created a system to monitor patients.

The indoor growing of cannabis does not have a season. Each grower packs in as many plants as possible in a batch, staggers the growth cycle of each batch based on harvesting, and turns over the greenhouse many times each year with as much yield as possible. In the same way that there is a cottage food industry, there is a cottage cannabis industry, and they can grow any way they want and however much they want under state licensure. Outside the cannabis cottage industry, cannabis businesses are pushing through as much cannabis as they can to try to meet the demand. Along with this difference from the food industry and because cannabis is illegal at the federal level, all legal cannabis must be grown in the state where it is processed and where edibles are manufactured. Cannabis, the plant, and cannabis products may *not* cross state lines in any form of transportation. A vial of concentrate cannot be shipped overnight from another state for you to run a product development batch. A bucket of cannabis butter cannot be placed in an ice chest and driven to you from the next state over. And you certainly cannot export or import cannabis plants or products. I don't know if cannabis or hemp seeds can be shipped across state lines; I hope so.

Cannabis plants for indoor growing usually start from clippings instead of seeds. The greenhouse will hold a section for the big, beautiful plants called "mothers." From the mothers, clippings are taken to start another batch. Clippings will be handled with tender loving care, under their own grow light, and with daily watering and observation, until grown enough to move to their first pot of soil. The big deal in cannabis growing is that the flower, the desired bud structure, is harvested from female plants. By starting with a clipping from a female plant, the grower is ensured to grow a female plant. Thus the name of this group of plants is mother plants, not father.

Marijuana grown for research

The Drug Enforcement Administration (DEA) recognizes the University of Mississippi as a registered grower of marijuana for research purposes [4]. Following the signing of the 2018 Farm Bill, DEA announced that marijuana production will grow (pun intended) by increasing the amount of marijuana grown for research purposes from 2450 to 3200 kg in 2020 [5]. Obtaining hemp or cannabis plant material and cannabis products for research purposes has been difficult. It was reported during the 2019 FDA public meeting [6] that legal cannabis products could be purchased by an individual for consumption, but a university researcher could not purchase cannabis for research projects because marijuana is illegal at the federal level. When the DEA deschedules marijuana off the Schedule I drug list, the door will open to the United States for medical, agricultural, and industry-applied research.

Postharvest steps for cannabis products

One way for me to introduce cannabis products is to follow the licensing (Fig. 2.1). Within a state that has legalized medical marijuana or recreational cannabis, there will be licenses for different steps from seed to sale. Each license will fall under the jurisdiction of a department in the state. Taxation aligns with the Department of Revenue. Other departments may include the departments of agriculture, health, enforcement, or licensing. In some states, marijuana departments have been created with the new state regulations. One thing is for sure, *every state is different*. Every state is handling the marijuana and cannabis laws in a different way. What is a food scientist like me to do? First, I live in Wisconsin, and we only have hemp-derived CBD products like every other state and district in the United States. I do purchase and use a CBD tincture (oil-based CBD dispensed with an eye dropper) for my autoimmune disorder. If I lived in one of my neighboring states, I would research the state code for marijuana. Minnesota has medical marijuana. In 2018, Michigan was the 10th state to legalize recreational cannabis after already having medical marijuana. Illinois has medical marijuana and is in the process of writing and implementing recreational cannabis laws as the 11th state.

Marijuana licensing in Colorado

Because I am more familiar with Colorado, Colorado has a relatively long history with licensing, and Colorado has recreational cannabis, I am going to use the Colorado licensure to explain [7]. In the law, M is for medical and R is for recreational. The R 200 Series-Licensing and Interests details the license fees and frequency of payment. There is a license for a Retail Marijuana Establishment Operator, and the requirements of the operator are found in the R 1700 Series-Retail Marijuana Establishment Operators. There are license fees for individuals for Associated Key License, Key License, or Occupational License based on the person's access to cannabis inventory, Limited Access Areas, and/or the Inventory Tracking System. Transferring this concept to the food industry would require food facility managers, supervisors, and employees working in supply chain management, shipping, or logistics to pay a license fee. If the cannabis company is operating in a state with vertical integration, there will be multiple fees to cover each step.

Retail marijuana testing facility

There is a license for a Retail Marijuana Testing Facility, with the requirements found in the R 700 Series-Retail Marijuana Testing Facilities. All testing laboratory facilities and personnel must follow the requirements for the R 1500 Series-Retail Marijuana Testing Program, as appropriate to their position. In Colorado, cannabis laboratories must be ISO 17025 certified to the cannabis tests the laboratory will perform.

Retail marijuana transporter

There is a license for Retail Marijuana Transporter, with the requirements found in the R 1600 Series-Retail Marijuana Transporters. All facilities and personnel must follow the additional requirements for the R 800 Series-Transport and Storage.

Retail marijuana cultivation facility

There is a license for a Retail Marijuana Cultivation Facility, with the requirements found in the R 500 Series-Retail Marijuana Cultivation Facilities. There is an Expanded Production Management Fee based on the number of plants over 1800.

Retail marijuana products manufacturing

There is a license for Retail Marijuana Products Manufacturing, with the requirements found in the R 600 Series-Retail Marijuana Products Manufacturing Facilities. Separately, there are requirements in the R 1000-1 Series-Labeling, Packaging, and Product Safety. Manufacturing includes the production of concentrate, topicals, tinctures, or Edible Retail Marijuana Product.

Hemp-derived and cannabis concentrate

Retail Marijuana Concentrate is extracted from the cannabis flower or trim to remove the solid plant material and concentrate the THC and CBD. Cannabis extraction facilities typically employ one of the following extraction processes:

- water-based Retail Marijuana Concentrate with potable water,
- food-based Retail Marijuana Concentrate with propylene glycol or glycerin,
- heat/pressure-based Retail Marijuana Concentrate,
- solvent-based Retail Marijuana Concentrate is common with butane, ethanol, and super critical carbon dioxide (CO_2) extraction. Colorado also allows propane, isopropanol, acetone, heptane, and pentane; all other solvents are prohibited.

Concentrate as an ingredient in an Edible Retail Marijuana Product should be made without solvents or with food-grade solvents such as CO_2, or food-grade ethanol, propylene glycol, or glycerin. The potential chemical hazard of the solvent should be addressed in the facility's hazard analysis (see Chapter 6, The Hazard Analysis).

Heat/pressure-based extraction [8]

Heat/pressure-based Retail Marijuana Concentrate production does not use solvents and produces kief, hash or hashish, or rosin from the trichomes of the flower. Each has its role in cannabis use. Kief is the sticky resin produced by the trichomes, rich in cannabinoids, and separated from the plant material. In cannabis growing and processing, removing the kief from surfaces is a cleaning challenge. Kief is separated from the flower structure by grinding and sieving to produce a powder. The powder is added to cannabis products to increase potency. Hash is the trichome structure of the flower with resin that is separated from the rest of the plant. Hash contains the plant material of the trichomes. Starting with the flower, hash, or kief, rosin is extracted by applying heat and pressure. The extraction process produces a concentrate that is rich in cannabinoids and terpenes and may be comparable to a concentrate from a solvent-based extraction.

Winterization [9]

The process of winterization is conducted on an extract for the purpose of removing waxes and concentrating the cannabinoids. The extract is heated with ethanol and cooled. Waxes and other material separate out during cooling and are physically separated from the cannabinoids dissolved in the liquid portion. The ethanol is removed by evaporation. Winterization can be performed on an extract or after the extract has gone through decarboxylation.

Decarboxylation [10,11]

The biochemically active THC and CBD are not found in the raw or dried cannabis flower. Cannabinoids in the natural state contain a carboxyl group and are present in the acidic form such as THCA and cannabidiolic acid. Removal of the carboxyl group is decarboxylation, and the chemical reaction is achieved at a high temperature. The process of smoking creates the required high temperature for decarboxylation, and biochemically active cannabinoids are inhaled. Hemp or cannabis plant material may be held at a given temperature for decarboxylation before extraction at a cannabis extraction facility. Decarboxylation of an extract is achieved with the control of temperature and time. Decarboxylation can be done on a cannabis extract directly or after winterization. Conversely, the decarboxylated extract can next go to winterization. Use of the nonpolar solvent butane for extraction is efficient in capturing and concentrating the decarboxylated cannabinoids. A sample of extract must be sent for testing of THC, CBD, and other chemicals to determine the safety, yield, and potency before going into a topical or edible. Cannabis-infused topicals and edibles are formulated with extracts containing the biochemically active form of THC and CBD.

Distillation

Cannabinoid isolates can be made by distilling an extract with the control of temperature. Vapors condense and different fractions are collected at the beginning, middle, or end of distillation, based on the properties of the desired isolates. The isolates are sold at a premium price to manufacturers and marketed to consumers looking for the medical benefits of certain cannabinoids.

Cannabis-infused edibles

Edible Retail Marijuana Products commonly include:

- chocolates
- soft confections, e.g., gummies
- hard confections, e.g., lozenges
- baked goods
- pressed pills and capsules
- liquid Edible Retail Marijuana Products, e.g., tinctures

As a food scientist, when I say that cannabis-infused edibles are shelf-stable, what is the property to monitor? Water activity (a_w, i.e., *a sub w*)! At this point in history, most edibles have a very low water activity, below 0.8 for safety or much lower to prevent spoilage. This makes the edible shelf-stable, not requiring refrigeration or freezing. Food microbiologists like the characteristic of low water activity to prevent pathogen growth. Food chemists like low water activity to maintain freshness and flavor profiles; the cannabis flavor profile develops from terpenes and will be preserved in low-water-activity flower or products. Some state code and the Cannabis Act in Canada require edibles to have a low water activity [12]. In addition to microbiology, chemistry, and regulation, there is another reason cannabis-infused edibles should have low water activity, and that is because the cannabis industry lives under a sample, hold, and release program.

Starting with the dried flower, including concentrate, and for finished product, there is a sample and hold program at the end of each licensing step (Fig. 2.1). States require rigorous sampling and testing programs, and cannabis companies are captives of the cannabis laboratory, their fees, and their turnaround time. At each step the cannabis material or product will not be released for sale until passing required tests for contaminants and the labeling of potency of THC and CBD. Without knowing the potency test results, packaging cannot be completed. Primary packaging of individual units may be applied, but each unit will have a sticker applied with the potency declaration. For example, at a Colorado dispensary, I purchased cannabis-infused chocolate. The packaging was beautiful, elegant. It was a rectangular box about 1 × 1 × 4 inches. On the outside of the box was a plain white sticker, bigger than an address label with the required label information, including potency. The white sticker could have easily been applied after the box was filled and held. I wonder about single cookies or bars that are individually wrapped in flexible packaging. It would be difficult to apply a sticker to the filled package compared with applying the sticker to the package first and then adding the cookie. Another truth is that there is a lot of manual labor in the cannabis-infused edible industry. There are great opportunities in the cannabis industry for packaging and labeling expertise; see Chapter 10 for more discussion.

Another difference between the food industry and cannabis industry is the value of the product and the measures taken to not waste or discard any. Have you ever used a mold and made individual pieces of chocolate? I think I did that once. It is a lot of physical labor and attention to detail. Now imagine that the cannabis chef is so skilled that he or she fills each mold cup with the exact same weight, does not drip any product between cups, and uses the entire batch of cannabis-infused chocolate without waste. It could happen. Cannabis-infused edibles are gourmet, high-end, and really expensive to make and to purchase. Wasting cannabis-infused edible work in progress would be like buying an expensive bottle of wine, pouring it, and then spilling the glass of wine!

Retail marijuana store

There is a license for Retail Marijuana Store, with the requirements found in the R 400 Series-Retail Marijuana Store. In addition, there are Colorado regulations for medical marijuana facilities, medical research facilities, and pesticide manufacturers.

In summary, following the state code for licensing will follow the path of the cannabis industry. Chapter 12, Jobs in the Cannabis Industry, dives into the roles and responsibilities of employees in the cannabis industry (Box 2.1).

Box 2.1 Real people in cannabis

Sanford Wolgel, Ph.D, Principal, KCNM Consulting, Illinois.

Time in the cannabis industry: 2 years in the current position and 6 years total in the cannabis industry.

What drives me in my job: The exciting and evolving science of all things cannabis.

References

[1] The name of Cannabis: a short guide for nonbotanists. https://www.ncbi.nlm.nih.gov/pmc/articles/PMC5531363/. [Accessed 09.15.19].

[2] The anatomy of a cannabis plant, and its lifecycle. https://www.visualcapitalist.com/anatomy-cannabis-plant/. [Accessed 09.11.19].

[3] Food and Drug Administration, Public Meeting: Public meeting on a new era of smarter food safety. https://www.fda.gov/food/workshops-meetings-webinars-food-and-dietary-supplements/public-meeting-new-era-smarter-food-safety-10212019-10212019. [Accessed 09.28.19].

[4] Marijuana research. https://pharmacy.olemiss.edu/marijuana/. [Accessed 09.28.19].

[5] DEA increasing Marijuana production by 30% to boost research. https://grizzle.com/dea-marijuana-research-30/. [Accessed 09.15.19].

[6] Food and Drug Administration. Public hearing: scientific data and information about products containing cannabis or cannabis-derived compounds. May 31, 2019. https://www.fda.gov/news-events/fda-meetings-conferences-and-workshops/scientific-data-and-information-about-products-containing-cannabis-or-cannabis-derived-compounds. [Accessed 09.01.19].

[7] Colorado department of revenue, marijuana enforcement division. Colorado marijuana laws and regulations. Charlottesville, Virginia: LexisNexis Matthew Bender & Company, Inc.; 2019.

[8] Cannabis solvent-free extraction. https://synergistictechassociates.com/extraction/solvent-free/. [Accessed 09.11.19].

[9] Winterization of cannabis extracts. https://synergistictechassociates.com/extraction/winterization/. [Accessed 09.11.19].

[10] Decarboxylation. https://synergistictechassociates.com/extraction/decarboxylation/. [Accessed 09.11.19].
[11] Iffland K, Carus M, Grotenhermen F. Decarboxylation of Tetrahydrocannabinolic acid (THCA) to active THC. Germany: European Industrial Hemp Association; 2016.
[12] Cannabis. https://www.inspection.gc.ca/plants/cannabis/eng/1531250234189/1531251514243. [Accessed 09.28.19].

CHAPTER

Begin with the end in mind®: the dispensary

3

The dispensary

The cannabis dispensary serves as the retail operation for the sale of cannabis products in the same way that convenience stores and supermarkets sell food. At this time, most cannabis products are shelf-stable, so there is no refrigeration or freezer units required at a dispensary. Because there are minimum age requirements and taxation, a dispensary may be more closely aligned with a liquor store than with a grocery store. In 33 states with medical marijuana, a state-issued medical card is required, so those states are not on my list for *field trips*; I can only enter dispensaries in states with recreational cannabis. Field trips are what I call visits to dispensaries in states with legal recreational cannabis. For my gigs of training or consulting or when I attend a professional conference, I travel outside my home state of Wisconsin to the 48 contiguous states. Ten states have legalized recreational cannabis, and Illinois will be the 11th state. Starting January 2020, my visits from Wisconsin to Illinois may include a field trip. As a nonresident of Illinois, I can purchase one-half ounce of flower or purchase edibles with up to 250 mg of Δ^9-tetrahydrocannabinol (THC) [1]. Edibles will start being sold in Canada in December 2019, so I am hoping for a work trip to Canada soon.

Washington

Visiting a dispensary is a strange experience and each dispensary has been unique. My first field trip was in Seattle. I was there to deliver training. A Green Bay friend got me in touch with her sister in Seattle. I met the sister and after dinner and drinks, we stopped at a dispensary. It was about 9 p.m. and dark out; the dispensary was well-lit like a pharmacy; even the display cases were well-lit. It felt chic and modern. A man was stationed at the door on a stool and checked our ID to confirm we were 21 years or older. No problem, I am older! Shopping for a cannabis product was like me shopping at Tiffany's. I could look but not touch. Typically, a salesperson, called a budtender, works with one customer at a time, in my case answering lots of questions, and making the sale. In my limited experience, budtenders are educational

Food Safety Lessons for Cannabis-Infused Edibles. https://doi.org/10.1016/B978-0-12-819512-3.00003-1

ambassadors and never push sales. After the sale, the order is filled from a back room and handed to you in a lunch bag.

A word about budtenders. Budtenders are akin to bartenders in that they take your order and see that you are served. They deal with all facets of the population and tend to show great care and tenderness to people. You can expect your budtender to be highly knowledgeable about all the products in the dispensary. They can talk about what to expect from a given product or a given ratio of THC to cannabidiol (CBD). Budtenders tend to listen well and strive to educate their customers.

Having never smoked or consumed edibles, I had no idea what I was looking at in the display case, so the sister explained some of the products. Behind the counter was a display of blunts, i.e., joints rolled in brown tobacco paper and sold individually. I now know that many dispensaries will roll their own joints and blunts for sale using the nugs and pieces that fall off in the jars of flower. In the display case were individual joints in sealed, glass test tubes. I randomly chose a package of gummies and a chocolate bar. Some thought of the CBD and THC content went into the decision. I went to pay for my products and was told (reminded) they only take cash, but there is an ATM in the corner. Most dispensaries can only accept cash because banking is controlled at the federal level and cannabis is illegal at the federal level, so it's illegal for banks to operate with illegal businesses. It's complicated. After the dispensary, I was dropped off at my hotel and prepared to teach the next day with no consumption of edibles (Box 3.1).

Box 3.1 I'm just curious, can you take cannabis-infused edibles home on the plane?

The answer to that question is the same answer to this question. Are you willing to risk jail for your gummies? No, you cannot.

The next evening, I was too afraid to carry the edibles with me, as I met a family member for dinner. I told the story of going to my first dispensary and how the sister told me that the Transportation Security Administration (TSA) doesn't care anymore about cannabis products and I can pack them in my checked bag and take them home. The response was, "No! I do not want you to go to jail! I do not want to see you on the news!" Thus started my education in contradiction in the cannabis industry. Whatever you read or hear in the cannabis industry may not be true, unless you are reading state legislation. That was my last evening in Seattle, and there was no consumption of edibles. I left both the gummies and chocolate for housekeeping.

Florida

I was privileged to speak about food safety in the cannabis track at the November 2018 Food Safety Consortium. There I met a cannabis chef working in Florida. We communicated over the months following the consortium, and I had a trip to Gainesville, Florida, in February of 2019. She was working in a cannabis kitchen start-up, and the kitchen was still under construction, so I was unable to visit her there, but we did meet for dinner. The next day she was able to get me into a tour of one of the company's dispensaries, close to where I was staying. I met the dispensary manager there; Florida only has medical marijuana. It was really nice because just the two of us started out in the consultation room just off the lobby. The consultation room is where patients can meet with budtenders to privately discuss their condition, decide on CBD and THC that would best address the condition, and choose a product for delivery of the CBD and THC. I heard an excellent lesson in cannabis chemistry and asked a lot of questions. There are good people in the cannabis industry who love to help others as evidenced by this manager. Legally, it was impossible for me to purchase product because I did not have a Florida medical card.

California

Before Christmas 2018, I knew I was booked to deliver training in San Diego, California. My son was making me a grandmother in 2019, so I wanted a trip with him before the baby arrived, and I gave him a plane ticket to San Diego for Christmas. With my travel covered, it was a frugal trip for us. As soon as we had checked into the hotel, we took an Uber to a dispensary. We were staying downtown, and the dispensary was near the naval base. The location was fantastic, and it was a beautiful San Diego day.

We walked through the door only to be trapped in a small vestibule. The locked door to the dispensary was to the right, and we gave our IDs to the woman sitting behind the glass window. My son had flown with a paper copy of his driver's license but could not get in the dispensary without a laminated ID. I was in and asked for the manager, so I could ask lots of questions. Like a kid outside a candy shop, he looked through the bars on the window into the store, but couldn't see, so he FaceTimed me! I flipped the picture so he could order product. Funny story: he chose a strain of cannabis flower. The budtender asked if I wanted indoor or outdoor. I immediately answered, "Indoor," thinking that indoor meant it would have less smell! When I realized I had no idea what indoor and outdoor meant, I asked. Indoor is cannabis grown indoor, and outdoor is cannabis grown outdoor. I think I stayed with indoor, but I really had no idea of the ramifications only that the price was different. The dispensary was fast paced with at least five budtenders and two cash registers. I remembered to bring cash. It reminded me more of a liquor store in a seedy part of town that was small, crowded, and dirty. When I emerged from the dispensary, my son had befriended the security guard (Box 3.2).

Box 3.2 Buying flower [2]

As a scientist who is comfortable in the metric system, I was confused about the units for buying cannabis flower for smoking: *eighth*, *quarter*, *half*... of what? I taught my students to always label their units, and I didn't understand the unit. The dried flower has volume without much mass and is sold in a fraction of an ounce. With 1 oz weighing (i.e., *massing*) in at 28 g, divide by 8, and you get 3.5 g. The most common unit for buying flower is an *eighth* of an ounce, and you receive 3.5 g of flower when you purchase an *eighth*.

Colorado

Every field trip gets a story, but the best Colorado story is in Chapter 10. I had a consulting gig in Denver, Colorado working for a cannabis company. I was able to purchase cannabis-infused chocolate at a company dispensary during my time on the job! Life is funny. My budtender was awesome, answering all my questions. The dispensary was not flashy like the Seattle dispensary, but run with high efficiency and professionalism.

Michigan

Michigan legalized recreational cannabis in the fall of 2018. If you know US geography, you know that the Upper Peninsula (U.P.) of Michigan lies north of Wisconsin. From my home, it is a short 1-hour drive for me to reach the U.P. of Michigan. I took a field trip to Michigan in the summer of 2019. It was another fun field trip with my son on a beautiful summer day, so that part was awesome. Other parts of the experience were disappointing and best not immortalized in writing. At a dispensary, I heard the black market still operates at about half the price for the flower at the dispensary. The cost of running a dispensary adds licensing fees, taxes, utilities, rent, and employee wages. The cost of the product includes expensive packaging and laboratory testing.

Nevada

What happens in Vegas, stays in Vegas. I am geeked out for my trip to Las Vegas, Nevada, where recreational use is legal, and I will have the opportunity for a field trip to a dispensary. I will attend the annual MJ Business Conference, i.e., MJBizCon, at the Las Vegas Convention Center in December 2019 along with 35,000 other people! In the food industry the largest convention I have attended is of the Institute of Food Technologists, which is near 20,000 attendees. I simply cannot image an exhibit hall with over 1000 exhibitors. The numbers are staggering, and I am excited for the education, networking, and adventure. Note: You can read about my experience in the final chapter of this book.

Home delivery

An extension of the dispensary is home delivery. Patients who have difficulty leaving their home because of illness or injury can have cannabis products delivered by courier service. This option is not meant to be like pizza delivery. When you think about companies and pharmacies that ship prescription drugs to the homes of patients, courier delivery of cannabis products makes sense (Box 3.3).

Box 3.3 Myth: Cannabis companies are making money hand over fist

No! It is extremely expensive to operate a cannabis business. It is true that cannabis companies are bringing in huge revenues, mostly in the form of cash. The profit, if any, is low.

Taxation and licensing

Considering the expenses of running a cannabis company, a great burden is taxation. Every state writes their own tax code, and the variation is great [3].

I am not a tax expert, so the information in Table 3.1 may not include all the taxation in a state, and the states of Maine, Massachusetts, Michigan, and Vermont are actively writing tax code. Cities, municipalities and counties may impose additional taxes to the state sales tax. The split is almost equal, but most states do tax the sale of medical marijuana with state sales tax [4].

Table 3.1 Taxation of cannabis cultivators and customers (*September 2019*)[a,b] [3].

State	Tax paid by cultivators	Tax paid by customers: Excise tax	Tax paid by customers: State sales tax
Alaska	$50 per dry-weight ounce	None	No state sales tax
California [c]	$9.25 per dry-weight ounce	15%	7.25%
Colorado	15%	15%	2.9%
Maine[d]	In the process of establishing a legal market	To be set	10%
Massachusetts	None	10.75%	6.25%
Michigan[e]	None	10%	6%
Nevada	15%	10%	4.6%
Oregon	None	17%	No state sales tax
Vermont	In the process of establishing a legal market	To be set	6%
Washington	None	37%	6.5%
District of Columbia[f]	There is no retail sale of recreational cannabis products	None	None on medical marijuana

Websites were accessed August 09, 2019.

[a] *https://www.nolo.com/legal-encyclopedia/state-taxes-on-recreational-marijuana.html.*

[b] *https://www.foxrothschild.com/content/uploads/2019/06/Cannabis-Tax-E-book-June-v22019.pdf.*

[c] *http://www.cdtfa.ca.gov/taxes-and-fees/tax-rates-stfd.htm.*

[d] *https://www.pressherald.com/2019/06/27/mills-signs-marijuana-regulations-into-law/.*

[e] *https://www.michigan.gov/documents/marijuana/MI_Regulation_and_Taxation_of_Marijuana_Act_Talking_Points_1.14.19_648379_7.docx.*

[f] *https://weedmaps.com/learn/laws-and-regulations/district-of-columbia/.*

Here is a case study on California for the collection and payment of taxes. Taxes are paid by cultivators on dry-weight cannabis flower ($9.25 per ounce), dry weight cannabis leaves ($2.75 per ounce), or unprocessed, fresh cannabis plant ($1.29 per ounce) [5]. The cultivator pays to either a manufacturer or a distributor on the first sale of the cannabis material. If the cultivator sells to a manufacturer, the manufacturer will pay the distributor. The definitions for *cultivator*, *manufacturer*, and *distributor* are defined by the California Tax Guide for Cannabis Businesses [6].

- *A cannabis cultivator is a person who is in engaged in the business of planting, growing, harvesting, drying, curing, grading or trimming cannabis. Even though the steps after harvesting look like processing steps, the steps of drying, curing, grading and trimming are commonly performed at the site of the cultivator.*
- *A processor license is a type of cultivation license issued by the California Department of Food and Agriculture. A processor performs trimming, drying, curing, grading, or packaging of cannabis and non-manufactured cannabis products. A processor both grows the cannabis and applies the next steps to the plant including packaging. At the first sale of the cannabis, the processor pays the cultivation tax.*
- *A cannabis distributor is a person who procures, sells, and/or transports cannabis between licensed cannabis businesses, such as a cultivator, manufacturer, or retailer. A microbusiness licensed to act as a distributor must comply with all the same requirements as a distributor.*
- *A cannabis manufacturer is a person who produces or prepares cannabis or cannabis products at a fixed location, that packages or repackages cannabis or cannabis products, or labels or relabels its container.*
- *A cannabis retailer is a person who sells cannabis and/or cannabis products directly to a consumer.*

The *retailer* pays the excise tax to the distributor for the cannabis product. All cannabis taxes in California are collected by the distributor and paid to the California Department of Tax and Fee Administration by the distributor. In California, the dispensary customer or cannabis product consumer pays the cannabis *retailer* both a 15% excise tax and a 7.25% state sales tax for the purchase. The retail sale of immature plants, clones, and seeds has both the excise tax and state sales tax applied. The *retailer* pays the state sales tax directly to the state as do all retail businesses. The black market is crushing the California cannabis companies. In early 2019 a bill was introduced to reduce *the state's excise tax from 15% to 11% for a period of 3 years and remove the cultivation tax on growers until 2022* [7]. In May 2019, Assembly Bill 286, dubbed the Temporary Cannabis Tax Reduction bill, failed to clear the Assembly Appropriations Committee and won't be seen for a vote on the Assembly Floor [8].

Continuing with the case study on California, Table. 3.2 [9] shows permit and licensing requirements for nine types of businesses. Remember that only the distributor pays cannabis taxes to the state; a distributor or microbusiness operating as a distributor must apply for a cannabis tax permit. Eight of the nine types of businesses must apply for a seller's permit; a cannabis testing laboratory applies for a seller's permit if it sells laboratory supplies. Note that in California, there are three government agencies licensing different types of businesses:

- California Department of Food and Agriculture;
- California Department of Consumer Affairs, Bureau of Cannabis Control;
- California Department of Public Health.

Table 3.2 Licensing of the cannabis industry in California [9]

Type of Business	Permit(s) Required	Cannabis License Required	Cannabis License Issued by the California Department of
Collective/cooperative (retailer)	Seller's permit	Retail license	Consumer Affairs, Bureau of Cannabis Control
Cultivator	Seller's permit	Cultivator license	Food and Agriculture
Distributor	Seller's permit* Cannabis tax permit	Distributor license	Consumer Affairs, Bureau of Cannabis Control
Manufacturer	Seller's permit	Manufacturer license	Public Health
Microbusiness (combined cultivation, manufacturing, distribution and retail)	Seller's permit Cannabis tax permit**	Microbusiness license	Consumer Affairs, Bureau of Cannabis Control
Nursery	Seller's permit	Cultivator license	Food and Agriculture
Processor	Seller's permit	Cultivator license	Food and Agriculture
Dispensary (retailer)	Seller's permit	Retail license	Consumer Affairs, Bureau of Cannabis Control
Testing laboratory	Seller's permit***	Testing laboratory license	Consumer Affairs, Bureau of Cannabis Control

**A seller's permit is required for the sale of cannabis or non-cannabis products.*
*** A cannabis tax permit is required if the microbusiness is an authorized distributor.*
**** A seller's permit is required for the sale of non-cannabis products such as sampling supplies.*

CDPH LICENSES FOR CANNABIS MANUFACTURING

LICENSE TYPE	MANUFACTURING ACTIVITY	NOTES
TYPE 7	For manufacturers using volatile solvents, such as butane, hexane, or propane	Type 7 licensees may also conduct Type 6, N or P activities
TYPE 6	For manufacturers using nonvolatile solvents, such as carbon dioxide, ethanol, water, butter or oil, or performing extraction using mechanical methods	Type 6 licensees may also conduct Type N or P activities
TYPE N	For manufacturers performing infusion	Type N licensees may also conduct Type P activities
TYPE P	For manufacturers only packaging or labeling cannabis products	Type P licensees may only conduct Type P activities
TYPE S	For manufacturers operating on a registered shared-use facility	The shared-use manufacturing facility must be approved before Type S licensees may submit license applications

FIGURE 3.1

California Department of Public Health cannabis manufacturing licenses [10].

Expect similar division of licensing and enforcement in all states, but it is confusing and inconsistent, as each state writes its own code. Cities, municipalities, and counties may impose additional licensing. Within the California Department of Public Health, there are five licenses to manage (Fig. 3.1) [10]. Cannabis manufacturing includes the steps of volatile or nonvolatile solvent extraction, infusion, and packaging or labeling, and a facility with mixed processing is called a shared-use facility. Any facility can package and label product, but a Type P license is for a facility that only packages or labels. A separate license is required for every facility owned by a company.

The cannabis industry does not benefit from federal programs such as tax deductions and banking. Some credit unions or local banks have financed cannabis businesses, but the cannabis industry mostly operates as a cash-only business. Cannabis products are expensive. When the time comes to pay quarterly sales tax or excise tax, a business must follow procedures for security and payment. In California, if a business is going to pay in cash, the business must seek out the local office to obtain a No Cash Exemption Request form, and approval of the request may exclude the business from paying with electronic funds transfer. After approval, the business must make an appointment at least 21 days in advance. If not approved and if taxes are not paid electronically, there is a 10% penalty.

References

[1] Illinois weed legalization guide: will your city or town sell marijuana? https://abc7chicago.com/business/will-your-city-or-town-sell-recreational-marijuana/5553943/. [Accessed 09.28.19].

[2] How much is an eighth? https://weedmaps.com/learn/products-and-how-to-consume/how-much-is-an-eighth/. [Accessed 09.15.19].

[3] How high are taxes on recreational marijuana in your state? https://taxfoundation.org/2019-recreational-marijuana-taxes/. [Accessed 09.11.19].

[4] Bookbinder MS, Millis JD. Cannabis industry state tax guide. 2019. www.foxrothschild.com.

[5] California cannabis taxes. http://www.cdtfa.ca.gov/taxes-and-fees/tax-rates-stfd.htm. [Accessed 09.11.19].

[6] California tax guide for cannabis businesses. http://www.cdtfa.ca.gov/industry/cannabis.htm. [Accessed 09.08.19].

[7] California's ridiculous marijuana tax calls for drastic action. https://www.fool.com/investing/2019/02/03/californias-ridiculous-marijuana-tax-calls-for-dra.aspx. [Accessed 09.11.19].

[8] California lawmakers nix temporary marijuana tax cut. https://www.kpbs.org/news/2019/may/17/california-lawmakers-nix-temporary-marijuana-tax-c/). [Accessed 09.11.19].

[9] Tax guide for cannabis businesses: getting started. http://www.cdtfa.ca.gov/industry/cannabis.htm#Started. [Accessed 09.11.19].

[10] California cannabis manufacturing licenses. https://www.cdph.ca.gov/Programs/CEH/DFDCS/MCSB/CDPH%20Document%20Library/LicenseTypes.pdf. [Accessed 09.10.19].

CHAPTER

Good manufacturing practice compliance is not optional

4

Introduction to good manufacturing practices

What are GMPs?

Good Manufacturing Practices (GMPs) are facility-specific programs, procedures, and policies followed by employees and visitors. GMPs include employee training, personal hygiene and the wearing of personal protective equipment (PPE), pest control management, the potability of water, the maintenance of the grounds and facility, and sanitation of utensils, equipment, and rooms.

GMPs are quality management programs. Quality management programs are implemented by owners and managers in the running of a business. They include hiring practices, procurement of supplies, sales contracts, payroll, worker safety, and others. With each quality management program, managers and employees follow the procedures. For example, if paper needs to be ordered for the photocopier at work, there probably is a request form to complete, a vendor list, the purchase order from an office supplies budget, and placing the order. If the procedure is followed correctly, a box of paper shows up at the photocopier. If GMPs are followed correctly, the product is more likely to be a safe product. Conversely, if GMPs are not in place or followed, the product is likely to be unsafe. Quality management programs address unique business needs to keep the business operating. GMPs address the specific business need to produce a safe product.

Quality management programs are often called prerequisite programs in the food industry. A prerequisite is something that must come first, before taking the next step. In other words, if employees are not hired, supplies not purchased, and sales contracts not signed, there is no business and no reason to make the product. The prerequisite programs support the operation of the business. In the realm of product safety, if GMPs are not implemented and followed, the potential exists for an unsafe product, and both the consumers and business may suffer.

Upkeep of the facility

GMPs are a little tricky to define because they include features of the facility and management of the facility that most of the time do not directly affect product safety. Facility owners and managers are required to follow GMPs for the upkeep of the

Food Safety Lessons for Cannabis-Infused Edibles. https://doi.org/10.1016/B978-0-12-819512-3.00004-3

facility. For example, product safety is not directly affected by the frequency of mowing the lawn, cleaning of the toilets in the employee bathrooms, or repair of the roof as long as these GMPs are getting done. However, if the GMPs associated with these activities are neglected, product safety may suffer. If the lawn and grounds are not maintained, there could be potential harborage of pests which make their way into the facility. A pest management program is a GMP. If the toilets are never cleaned, there can be the spread of pathogens onto employees or through aerosols generated in flushing. A master cleaning schedule is a GMP that details the frequency of cleaning and sanitizing of all parts of a facility. Water from roof leaks can bring pathogens found on top of the roof inside the facility. Management is responsible for the upkeep of a facility and the integrity of the roof, doors, windows, and ventilation units as GMPs.

GMPs must be implemented regardless of the size or type of facility. Consider a warehouse for the holding of food that does no labeling or repacking and all the food is at ambient temperature with no refrigerated or frozen storage. What could possibly be a hazard at this warehouse? Rodents, leaky roof, damaged cases, and spills. While the direct control of a hazard is not addressed with GMPs, without GMPs, a hazard may emerge. If GMPs are followed, there will be no hazard to control at the warehouse. The warehouse facility still must follow GMPs for grounds control, building maintenance, pest control, and sanitation. In other words, every food facility must follow GMPs. There are no exceptions.

Training of employees and documentation of procedures

Write what you are going to do, do it, and document that you did it. GMPs have associated Standard Operating Procedures (SOPs) which serve as both documentation of procedures and employee training documents on those procedures. All employees are trained in some GMPs like proper handwashing. Other GMPs are specific to the position such as trash removal or changing of filters in ventilation units. In some ways, GMPs are the result of explaining "this is how we do things here."

Food and Drug Administration authority

The cannabis industry is not provided with GMPs from FDA because of the lack of federal oversight. The cannabis industry can look to the regulations for GMPs found in separate FDA rules for pharmaceuticals, dietary supplements, and food. Because GMPs are found in FDA rules, GMPs are utterly required and are not optional for facility management of cannabis-infused edibles. Congress gives FDA the oversight and authority to enforce rules. Changing a rule literally takes an act of Congress. In the signing of the Food Safety Modernization Act in 2011, FDA was given the authority to initiate a recall as a tool in its enforcement toolbox. Most recalls are initiated by companies working in cooperation with the FDA. In addition to recall authority, FDA has the authority to pull a facility's registration causing the facility to cease operations. FDA also uses the tools of warning letters requiring a response from the facility, federal marshal seizure of product, and bringing owners or managers to court and pressing charges. The FDA hands severe cases over to the Department of Justice as appropriate.

It is illegal to offer adulterated or misbranded food into interstate commerce. Adulteration means the food has the potential to cause illness or injury due to the presence of a biological, chemical, or physical hazard. FDA enforcement of GMPs is strict to prevent adulterated food. When food is misbranded, the contents do not align with the packaging and labeling. Examples are when the ingredient listing is wrong, when the layout of the label does not follow the rule, and when a package is underfilled compared to the labeled weight. In most cases, misbranded food is in the category of an economic problem, and the food would not cause illness or injury. When marijuana is descheduled, look to FDA for rules detailing GMPs for cannabis-infused edibles and know the authority given to FDA for enforcement.

Alignment of 21 CRF 111 and 21 CFR 117

FDA provides GMP requirements to dietary supplement manufacturers in 21 CFR 111 *Current Good Manufacturing Practice in Manufacturing, Packaging, Labeling, or Holding Operations for Dietary Supplements* [1] and to food companies in 21 CFR 117 *Current Good Manufacturing Practice, Hazard Analysis, and Risk-based Preventive Controls for Human Food* (PCHF) Subpart B Current Good Manufacturing Practice [2]. An alignment of the two regulations follows (Table 4.1). Pharmaceutical GMPs will not be addressed here due to the general agreement in the cannabis industry that cannabis-infused edibles will not be regulated as a drug but will be regulated as a food or dietary supplement.

At first glance of Table 4.1, the dietary supplement regulation appears more prescriptive on production and process control, and the food regulation looks heavy on sanitation. There is no reason to believe either industry is lacking in one area or another in the reality of day-to-day operations; it is just the presentation of the GMP requirements in the two regulations that differ on paper. There may be terms unique to each industry like Master Manufacturing Record and Batch Production Record for dietary supplements, but the work of following a recipe and recording the amount of each ingredient in a batch is the same requirement in both the dietary supplement and food industry, regardless of the name of the record.

Under FDA, there are rules separate from the PCHF rule [2] and GMPs for dietary supplements rule [1] for egg products, 100% juice, low-acid canned foods, and fish and seafood. If a cannabis company were to create a cannabis-infused edible from one of these food groups, the respective FDA rule must be followed. These additional rules from FDA do not include GMPs; GMPs are only found in 21 CFR 117 for the food industry. It does not matter what food product is made under FDA enforcement, GMPs are followed from 21 CFR 117. The previous version of GMPs was found in 21 CFR 110, and some may hear the old rule referenced. The old GMPs of 21 CFR 110 were replaced with the GMPs in 21 CFR 117, and companies should follow the latter rule.

Companies processing meat and poultry under USDA compliance have USDA rules and are not under FDA enforcement. Meat edibles are not in the US conversation of edibles, so it is prudent that cannabis companies in the United States stay away from meat and poultry processing and just stick to foods under FDA enforcement. In Canada, the Cannabis Act does allow for dried meat edibles with low water

Table 4.1 Alignment of 21 CFR 111 and 21 CFR 117 for current Good Manufacturing Practice.

Part 111—Current good manufacturing practice in manufacturing, packaging, labeling, or holding operations for dietary supplements	Part 117—Current good manufacturing practice, hazard analysis, and risk-based preventive controls for human food
Subpart A—General provisions	**Subpart A—General provisions**
	Subpart B—Current good manufacturing practice
Subpart B—Personnel	§ 117.10 Personnel
Subpart C—Physical plant and grounds	§ 117.20 Plant and grounds
	§ 117.35 Sanitary operations
	§ 117.37 Sanitary facilities and controls
Subpart D—Equipment and utensils	§ 117.40 Equipment and utensils
Subpart E—Requirement to establish a production and process control system	§ 117.80 Processes and controls
Subpart F—Production and process control system: requirements for quality control	
Subpart G—Production and process control system: requirements for components, packaging, and labels and for product that you receive for packaging or labeling as a dietary supplement	
Subpart H—Production and process control system: requirements for the master manufacturing record	
Subpart I—Production and process control system: requirements for the batch production record	
Subpart J—Production and process control system: requirements for laboratory operations	
Subpart K—Production and process control system: requirements for manufacturing operations	
Subpart L—Production and process control system: requirements for packaging and labeling operations	
Subpart M—Holding and distributing	§ 117.93 Warehousing and distribution
	§ 117.95 Holding and distribution of human food by-products for use as animal food
	§ 117.110 Defect action levels
Subpart N—Returned dietary supplements	
Subpart O—Product complaints	
Subpart P—Records and recordkeeping	**Subpart F—Requirements Applying to Records That Must Be Established and Maintained**

activity [3] like jerky, but I am hopeful that US edible manufacturers will not go down the meat path for both regulatory and safety reasons.

FDA GMP inspections and warning letters

FDA GMP inspections far outnumber preventive control inspections. It was reported at the 2019 annual conference for the International Association for Food Protection that for the FDA fiscal year 2018/2019, there had been 2483 FDA-led GMP inspections compared to 423 preventive controls inspections for compliance with the PCHF rule [4]. The importance of GMP to FDA is evidenced from how frequently GMP violations are cited in Form 483 given to facilities and warning letters. Dietary supplement manufacturers must comply with 21 CFR 111 *Current Good Manufacturing Practice in Manufacturing, Packaging, Labeling, or Holding Operations for Dietary Supplements* [1]. Here are excerpts of one warning letter sent to a dietary supplement manufacturer in violation of GMPs [5].

> *The U.S. Food & Drug Administration (FDA) conducted an inspection of your manufacturing facility located at 1730 Martin Luther King, Jr. Blvd, SE, Suite 6, Gainesville, GA 30,504 from March 11 – 15, 2019. The inspection found that your facility has significant violations of the Current Good Manufacturing Practice (CGMP) regulations for dietary supplements, Title 21, Code of Federal Regulations (CFR), Part 111 (21 CFR Part 111). These violations cause the dietary supplement products you manufacture to be adulterated within the meaning of section 402(g) (1) of the Federal Food, Drug, and Cosmetic Act (the Act) [21 U.S.C. § 342(g) (1)] in that the products have been prepared, packed, or held under conditions that do not meet CGMP requirements for dietary supplements. The introduction or delivery for introduction into interstate commerce of any food that is adulterated is a violation of section 301(a) of the Act.*
>
> 1. *You failed to prepare and follow a written master manufacturing record for each unique formulation of a dietary supplement that you manufacture, and for each batch size, to ensure uniformity in the finished batch from batch to batch, as required by 21 CFR 111.205(a).*
> 2. *You failed to establish specifications for any point, step, or stage in the manufacturing process where control is necessary to ensure the quality of the dietary supplement, as required by 21 CFR 111.70(a).*
> 3. *Your batch production record did not include complete information relating to the production and control of each batch and did not include all information required in a batch production record, as required by 21 CFR 111.255(b) and 111.260.*
> 4. *You did not hold components, dietary supplements, and labels under conditions to prevent their mix-up, contamination or deterioration as required by 21 CFR 111.455(c).*
> 5. *You failed to identify and quarantine returned dietary supplements until quality control personnel conducted a material review and made a disposition decision as required by 21 CFR 111.510.*

6. *You failed to hold reserve samples of dietary supplements using the same container-closure system in which the packaged and labeled dietary supplement was distributed as required by 21 CFR 111.83(b) (1).*
7. *You did not make and keep records of the written procedures for the packaging and labeling operations as required by 21 CFR 111.430(b).*
8. *You failed to establish and follow written procedures for holding and distributing operations, as required by 21 CFR 111.453.*

Under each violation quoted above, details of observations were given. Every warning letter from the FDA comes with the caveat that the letter was not meant to be all-inclusive of violations. Think of each observation as a citation. The company had 15 days to respond to the letter which is a typical timeframe for FDA. The response should address each observation and detail the company's plans to tackle each observation.

Implementation of GMPs

I see two main areas of focus for the cannabis industry for GMP implementation: documentation and training. Once facility management has identified and written programs, policies, and procedures for GMPs, it is critical to keep the GMPs alive through training. In this way, documentation and training work together symbiotically. The implementation of GMPs will get better with training, and the training will get better as written GMPs and records are continually improved.

Documentation

Standard operating procedures

GMP documentation includes management policy, SOPs, the records associated with SOPs, and the completed records collected in recordkeeping. As a consultant, I am often asked by start-up companies if I will help them by writing their SOPs. It is a tricky position to be in because I do not mind the work, and it is something I *can* do but *should not* do for a company. SOPs are written documents giving step-by-step directions for GMPs. The list is almost endless but should include pest control management, receiving, handwashing, trash disposal, all cleaning and sanitizing procedures, allergen management, and operation of equipment (Appendix 1). The person writing the SOP should have a good understanding of how to do the procedure, address why the procedure is important to be done as written, and know how the procedure fits into the operation of the facility and within the blueprint of the facility for location of the procedure.

If I am not the one writing SOPs, who is going to take time to write all these SOPs for GMPs? The employees doing the procedure! Who is better to write the SOP than the person responsible for the work? If management does get involved in writing SOPs, as is the usual case, the written SOP should be reviewed while observing the employee doing the procedure, revised, and then reviewed by the employee for completeness. Documentation remains under continuous improvement. Find your person who loves document management. As I said, I am often asked to write SOPs. As you can see, if I write an SOP for a company, I need to be onsite or have unlimited access to the employee doing the procedure for that employee to tell me the steps and me to type the steps. Doesn't it just make sense to have that employee type up the steps? If that person is not computer savvy, pair them with someone in the office who is. The person in the office will enjoy learning about the work out on the floor. Also, add pictures to your SOPs.

After an SOP is in good shape, it becomes a training document. Let us say you write an SOP for the inspection of the grounds around the facility. You may include a blueprint layout of the perimeter of the property and the buildings on the property. There will be step-by-step directions for what to observe and what to do. For example, litter that has blown in around fencing, landscaping, and buildings must be collected and discarded. Separate SOPs will be written for landscaping and grounds maintenance. The lawn should be mowed and trimmed; trees and bushes should be trimmed at a frequency determined by the facility management.

What is the frequency of each SOP? Some facilities located in the city need to collect litter daily; some locations can collect litter weekly. On the food side of GMPs, there is no guidance from FDA on the frequency of GMPs. Frequency is completely up to the facility management and what is needed to make a safe product and maintain a sanitary facility. In many cases and along with the SOP, a bench sheet or record needs to be made and then retained. The record can be a handwritten hard copy or electronic. Do not make this harder than it has to be. A simple checklist is fine. Let us say the grounds inspection and trash collection are completed on a daily basis. Create a checklist of each step and have the employee completing the task initial and date the checklist. Have you seen the bathroom check sheet on the back of a bathroom door in a restaurant or other public location? If the property wants the public bathroom checked hourly, there is a place to date, initial and note the time the bathroom was checked. It is very easy to see the directions on the top and (1) was the bathroom checked hourly and (2) who was responsible. Keep it simple.

As the company grows and many employees are following the same SOP, there is the danger of procedure drift where steps are forgotten, or shortcuts are made. Supervisors should take the opportunity to observe employee procedures against the written SOP to verify procedures are being done as intended in the SOP.

The topic of SOPs reminds me of when I worked at a cheese company as a quality control technician and discovered the bible of the dairy industry in the cupboard, *Standard Methods for the Examination of Dairy Products* (SMEDP). This was months after I was trained by being shown and told what to do. That style of training meets the needs of many employees, but my learning style of reading is strong. SMEDP explains the details of laboratory methods and standardizes the methods across the industry. I followed the written procedures in SMEDP along with the specifics of my work site for fulfilling the testing, gaining confidence that my work was accurate. Many cannabis employees from the growing operations to the dispensaries are researching and reading on their own time. They are all in. Share cannabis resources with employees. Point them in the right direction for resources of good information and support training at any opportunity. See Chapter 11, Training Needed in the Cannabis Industry, for resources both in and out of the cannabis industry.

Sanitation standard operating procedures

Sanitation Standard Operating Procedures (SSOPs) are critical to the overall facility cleanliness and in the prevention of allergen cross-contact and pathogen cross-contamination in a cannabis-infused edibles facility. Not only is thorough cleaning of food-contact surfaces required as a GMP and as noted in the FDA warning letters above, but when an SSOP is followed there must be adequate recordkeeping to prove that the sanitation crew followed the SSOP. Many companies fail at the recordkeeping step. A company decides how frequently to clean and sanitize and by chosen methods and writes the SSOP. The master sanitation schedule instructs on what parts of the facility and equipment are to be cleaned in what frequency and by which SSOPs. Records must be kept proving the company did what they say they do in the master sanitation schedule and its SSOPs. When I talk to clients about SOPs, I usually say, "written SOPs with associated records for documentation." An SOP is married to its documentation for recordkeeping.

GMP training

In the food industry, employees are asked if they would serve the product to their family. Is the product safe enough for their own family? GMP training teaches the procedures unique to every facility and provides depth in the reasoning and importance for the procedures with the goal of product safety. GMP training is a necessary burden to industry in the resources of time away from production or other tasks, development of training materials, and documentation of employee training. Managers wear many hats, and to deliver GMP training should be part of the job description for every supervisor and manager.

GMP training for employees

A facility can divide employee training among on-boarding of new employees, training to specific tasks or positions, and refresher training that is on-going.

On-boarding of new employees. It is common in the cannabis industry to hire new, temporary employees through temp agencies. In the food industry, it is common to hire seasonal help. My recommendation is to work with a temp or placement agency for the on-boarding process and include a GMP training program delivered by the agency, before the new employee walks in the door. Start-up and small businesses may not even have an office or space to conduct training onsite and moving this step to the agency will help. The on-boarding process should include all the GMP training an employee would need on day 1 of the job. Once the employee steps into the facility for their first day on the job, it is critical to maintain the culture of food safety and following GMPs. The culture becomes "this is what we do here." Everyone follows GMPs all the time. The new employee will apply the GMP training and see others doing the right thing too.

When hiring employees at a cannabis facility, state code varies for background checks. Some state code prohibits the hiring of a felon at a cannabis facility. Illinois comes on-board with the sale of recreational cannabis on January 1st, 2020, and cannabis companies need to hire more employees to meet the demand for their products. I recently heard that Illinois offers expungement fairs for ex-prisoners who qualify for expungement of their records and who wish to gain employment in the rapidly growing cannabis industry in Illinois. Expungement will erase their prison record and allow employment in the cannabis industry. Once hired, training specific to the facility begins.

GMP training is delivered to employees handling ingredients, packaging and the finished product; working with the product at any step in processing; and working with food-contact surfaces such as the sanitation crew and maintenance workers. Personally, I argue that every employee working at a cannabis-infused edible or food facility should receive GMP training. Employees change positions or leave the company, creating a hole in production that needs filling. When the company has a big order to fill, the administrative staff and managers may be pulled onto the line to get the order out. They need GMP training. Think of the situation of a staff member walking from the offices through the facility to shipping. That employee must follow GMPs. Every employee is working for a facility that makes an edible product and should know the GMP requirements of the company and its employees. Administrative staff should know the GMPs that visitors must follow and assist the visitors with preparing to enter the processing rooms. GMP training should be for everyone.

Training to specific tasks or positions. In addition to GMP training at onboarding of an employee, there is training to a position or task. As an employee gains experience and more responsibility, the employee must be trained to the corresponding GMPs for a new task or new position. This includes supervisor and manager training. As an employee moves into a supervisory position, their training record should reflect their skills. I recommend all training be recorded in the employee's personnel record.

One great part about having a written SOP is using the SOP for training. As an employee is trained to a new position or new task, the written SOP will be used to introduce the employee to their new responsibilities. Start with the employee reading the SOP and signing off on that step. Then the employee observes the procedure being done. When the employee does the new procedure, there is observation and correction, if needed, by a supervisor. Once the employee has mastered the new procedure and is working independently, the supervisor should observe the employee occasionally and verify the procedure is being done correctly as written in the SOP. This seems like a lot of work, but it does not have to be. Following steps in training builds confidence in both operators and supervisors that employees are correctly trained to a procedure as written and not going rogue with their own way of doing things.

Refresher training. To keep the GMPs fresh, there must be on-going training of GMPs. This can include day-to-day observations by supervisors, a monthly employee meeting, posting of signs and procedures, and so much more. With the prevalence of technology, could each employee receive a daily reminder by text on their personal cell phone? Could messages scroll on a monitor in the break area? Be creative and keep it fresh. After a while, don't we all just not see the sign in the restroom stating that hands must be washed before returning to work? Be creative in refresher training on GMPs.

Create a culture of quality and safety. GMP training is only the tip of the iceberg. The first goal is to train employees so that GMPs are implemented, followed, and monitored. Beyond the normal operations of a facility, employees should understand GMPs enough to recognize and report a problem. For example, after a heavy rain, the employee at shipping notices a roof leak. The employee should recognize the problem, contain the water, and report the leak for maintenance. When I work with clients in their facilities, I always point out issues with the integrity of the building when I see them. Most managers appreciate a fresh set of eyes on the facility and the perspective of someone new. Unfortunately, some managers need an outsider to say something is a problem for upper management to hear what the manager at the facility has been saying all along. I am happy to provide that service.

I am a human leak detector, and I often point out leaking pipes within a facility. It is a test of the facility culture once I point out the leak and if the leak is reported, contained, and repaired while I am still on-site. One time, I pointed out a leak on the side of a piece of equipment and continued to discuss the operation of the equipment. When we were ready to move on to the next step in the process, I turned around and bumped into a maintenance man fixing the leaky pipe! That is an example of an excellent culture of food safety. In one facility, I pointed out a dripping pipe to the quality manager as we walked through a dry room. The second day on site I pointed out the same dripping pipe. On the third day, we walked through, and the pipe was still dripping. In other situations, I notice a drip to the floor and follow overhead to locate the source of the leak from above. When I look back to the floor to see where the drip is located, I notice a pit in the floor. This drip has been dripping so long that the material of the flooring has eroded away. Why was the drip or leak never fixed? When facility management has adequately trained employees on GMPs, the employees understand the importance of following GMPs and fixing problems when they arise.

GMP training for visitors

A visitor may be a nonemployee, is not a temporary or seasonal hire, or may be an employee from a different facility within the same company. Visitors include persons coming in the facility for equipment repair or construction and customers, auditors, inspectors, and truck drivers. It is very common for a visitor to read a GMP policy for visitors and sign and date the document as their agreement to following the GMPs. A visitor should always be escorted by an employee trained in the GMP procedures for the type of visitor and to limit the visitor's access to restricted areas of the facility. The managers of the facility will provide additional verbal instructions to the visitors on GMPs.

Absolutes for training in the cannabis industry

Here is a short list of topics that must be covered in training employees in the cannabis industry. For a long list, see the Outline of GMPs for the cannabis industry (Appendix 1).

The immunocompromised consumer

In a food safety plan, the description of the consumer will often be *general population*. This is the case for most foods as most foods are made for consumption by the general population. The cannabis-infused edible is destined for a consumer with a preexisting condition and an already weakened immune system. In eleven states, patients with medical cards can shop for edibles at a dispensary that serves the adult population for recreational cannabis. Therefore, the cannabis-infused edible industry in 33 states with medical marijuana knows it is making edibles for consumers who are already ill. The fact that the cannabis industry knows the consumer is immunocompromised is the equivalent of the food industry selling product into assisted care living facilities, day care facilities, or hospital cafeterias. Food consumers with weakened immune systems include the very young, the old, pregnant women, and anyone with a medical condition in which the illness or drug therapy has weakened the immune system. The cannabis industry is not selling recreational cannabis to minors but may be selling to parents for treatment of a child's illness.

When a food or edible is made and sold to populations who have weakened immune systems, extra precautions are taken. The kill step in manufacturing is monitored, steps are taken to prevent cross-contamination after the kill step, equipment and utensils are thoroughly cleaned and sanitized, and employees exhibit the highest level of personal hygiene and wearing of sanitary PPE. PPE is designed to protect the product, not to protect the employee (Box 4.1).

> **Box 4.1 Cross-what?**
>
> *Cross-contamination* is the spread of a pathogen from one location to the finished product.
> *Cross-contact* is the transfer of an allergen from one location to an ingredient or finished product otherwise without the allergen.

Handwashing and personal protective equipment

Proper handwashing is one of the most important steps an employee can perform on the job. Do not take shortcuts, or the handwashing will not be effective in the prevention of cross-contact and cross-contamination. The handwash station must be maintained in a sanitary manner, be stocked with detergent and single-use towels, and provide warm water. Common observations cited by FDA that get companies in trouble are that the handwash sink was not dedicated, not accessed for handwashing, not stocked, or not maintained as sanitary. It is the responsibility of managers to model proper handwashing and provide refresher training of employees. Do not assume employees know how to properly wash their hands or know of pathogens or allergens transferred in handling. Employees must be trained in proper handwashing and use of PPE.

Along with proper handwashing in the prevention of cross-contact and cross-contamination is the wearing of PPE. From the top of the head to the tip of the toes, management policy dictates the required PPE for each workstation. The employee has the responsibility of changing PPE when it becomes unsanitary or ripped and of changing into fresh PPE when changing lines or positions where the product must be protected from cross-contact or cross-contamination. For example, a nut bar line completes its production for the day. Before an employee moves from the nut bar line to the non-nut bar line to finish their shift, the employee must completely degown, wash their hands, and don fresh PPE to prevent cross-contact of the nut allergen to the non-nut bars.

Traffic patterns

Restricted access and control of traffic patterns are means to prevent cross-contact as in the nut bar example above and prevent cross-contamination. Pathogens are potentially located outside the facility, in ingredients, on people, on wheeled equipment, in the air, and in drains. The traffic pattern of a trash bin on wheels is a good example. The trash bin is located next to the packaging of finished product. When full, the trash bin is wheeled out of the packaging room, through the warehouse, to the receiving dock, and emptied outside into the dumpster. The wheels of the trash bin have traveled through contaminated puddles and across potentially contaminated floors. The employee is wearing gloves that have touched doors, the trash bin, and the dumpster. The wheels and gloved hands spread the contamination along the path back to the packaging area. The employee continues packaging of finished product without changing gloves. Instead, the wheels of the trash bin should be sanitized and the employee should wash their hands and apply fresh PPE before returning to the line. Understanding the traffic patterns of employees and wheeled equipment such as carts, hand trucks, and forklifts is essential in the prevention of cross-contact and cross-contamination.

GMPs are only the beginning of safety for cannabis-infused edibles

For all cannabis products including flower for smoke, concentrate, and cannabis-infused edibles, GMPs are the foundation upon which product safety is built. GMPs will be required by FDA. The strong foundation of GMPs is the first step and is followed by a hazard analysis, preventive controls, and a written recall plan.

What is the food safety team called at a cannabis-infused edible facility?

I have not seen the crossover of the term *food safety team* to the cannabis industry, so my recommendation is to apply the analogous term *edible safety team.* The work of the edible safety team is the same as the work of a food safety team (Box 4.2). The term *safety team* implies the safety of employees at the job site which speaks to worker safety not the safety of the edible and would not be appropriate or analogous to the term food safety team. The edible safety team identifies hazards and verifies the hazards are controlled in the edible through the implementation of an edible safety plan. My recommendation is that the edible safety team has at least one member who is a Preventive Controls Qualified Individual as required in the food industry.

Box 4.2 What is a food safety team?

In the food industry, the *food safety team* is a mix of managers from the departments of Quality Assurance, Operations, Shipping and Receiving, Sanitation, Maintenance, Procurement, Human Resources, Regulatory Affairs, and any other department. The recommendation for team size is eight or less because there should be enough team members that a variety of perspectives are represented, yet not so many that the team stalemates in too much discussion without progressing in its work to conduct a hazard analysis (Chapter 6, The Hazard Analysis), implement a written food safety plan, and write a recall plan. A food safety team must have at least one member who is a Preventive Controls Qualified Individual (PCQI).

Elements of an edible safety plan

An edible safety plan includes a hazard analysis, preventive controls, and a written recall plan. Each is introduced here and explained in detail in following chapters.

Hazard analysis

The hazard analysis is a written document that notes potential hazards such as pesticides, heavy metals, residual solvents, and mycotoxins. The edible safety team analyzes every ingredient and process step for biological, chemical, or physical hazards. In contrast, GMPs do not specify hazards. For example, a pest management program as a GMP controls rodents and insects without naming the specific rodent or insect. The hazard analysis is very specific in naming the hazard. More information is in Chapter 6, The Hazard Analysis.

Preventive controls

Once a hazard is identified, the edible safety team will use resources to determine the risk of each hazard. Hazards with low risk may be managed with GMPs. Hazards with high risk are managed by one or more preventive controls from the categories of process, sanitation, supply chain, or allergen preventive controls. A preventive control from each category is not required. More information is in Chapter 7, Preventive Controls.

Process preventive control

Process preventive controls are the same as what might be the more familiar critical control points (CCPs) in a hazard analysis and critical control point (HACCP) plan. In requiring preventive controls, FDA moved beyond CCPs to include sanitation, supply chain, or allergen preventive controls. Process preventive controls typically use the parameters of time, temperature, acidity, or water activity for the killing of pathogens and prevention of growth of pathogens in process steps. Process preventive controls require validation of kill steps either from citing technical literature such as the FDA Food Code or by validating a kill step in-house.

In-house monitoring of process steps. No matter what means of process preventive control is used to kill and/or prevent growth of pathogens, data must be recorded for monitoring the process steps with a process preventive control. The employee recording the data must be trained and know if the process parameters have been met for killing or prevention of growth. A simple review of the record by a manager is an example of a verification activity for the process preventive control. First monitor the process, then verify control of the hazard.

Sanitation preventive control

Is *Salmonella* lurking in the processing or packaging rooms? A sanitation preventive control is required in the processing and packaging rooms for ready-to-eat (RTE) foods and can be expected to be required for cannabis-infused edibles by FDA. Following a kill step, the product is exposed to the environment until the primary packaging is sealed. A facility is required to effectively clean and sanitize the rooms, structures within the rooms, and equipment to remove and kill pathogens. RTE edibles have a high risk of being cross-contaminated with either *Salmonella* or *Listeria monocytogenes* as environmental pathogens and biological hazards, requiring a preventive control.

Common challenges in sanitation that get companies in trouble are inadequate hot water temperature at the exit from a hose for cleaning and sanitizing and improper use of a dedicated three-compartment sink.

Is sanitation a GMP or a preventive control?

Before the FDA PCHF rule [2] and under HACCP, the sanitation program for a facility was a GMP. The food industry has grown beyond HACCP to the implementation of a food safety plan including preventive controls, a procedure to control a specific hazard identified in the hazard analysis of a facility. For food safety plans, the conversation as to a sanitation SOP being associated with a GMP or a preventive control is a muddy one. The most important point for food safety is that the cleaning and sanitizing are getting done effectively; the conversation is around nomenclature and documentation. The training for PCQIs addresses the question of "Is this sanitation step a GMP or preventive control?" A PCQI is the person at a food facility to guide the decision-making with the food safety team around this question. For any RTE product, a sanitation preventive control is required to control the hazard of environmental pathogens. Along with that preventive control comes monitoring, verification, records, and correction or corrective action. A sanitation SOP for a GMP still needs the written SOP and its associated record but may not require verification and correction or corrective action. For example, if the employee breakroom does not get cleaned one day, it is not necessary to complete a corrective action form and do a root cause analysis as part of a corrective action, because the cleaning of the breakroom, while necessary, is not directly controlling a hazard.

Not all food and not all cannabis-infused edibles are RTE, but most edibles at this time are RTE. If an edible is not RTE, then the edible is ready to bake as with raw cookie dough, brownie batter, or pizza dough or is ready to cook as with frozen foods requiring cooking before consumption. Another example of edibles that are not RTE includes tea that will be steeped or cocoa mix that will be added to hot water or hot milk. Baking, cooking, or steeping are kill steps in the destruction of pathogens and applied after the sale of the product. Edibles with a kill step applied after the sale of the product would not require a sanitation preventive control, and sanitation of the facility and equipment may remain a GMP as is the case in the food industry.

Honestly, I am still wrapping my head around this difference between sanitation GMPs and sanitation preventive controls. I know that a hazard analysis must identify a hazard and that a sanitation preventive control may be implemented for that control. If there is no hazard in the hazard analysis that is controlled by a sanitation preventive control, there will be no sanitation preventive control. I understand the difference between GMP and sanitation preventive control in an academic way, but I know the application of the real work in a facility is muddy. It is no wonder the FDA catches errors in both the dietary supplement and food industries. A facility sanitarian can partner with technical support representatives at their chemical supply company to guide a food or an edible safety team and sanitation department in sanitation procedures, decision-making, and required documentation. The difference between a sanitation GMP and a sanitation preventive control is that a sanitation preventive control is specific to the control of the hazard of an environmental pathogen (Chapter 7). An environmental monitoring program (Chapter 8) verifies the effectiveness of the sanitation preventive control.

Written recall plan

Beyond current state regulation, a written recall plan for cannabis-infused edibles will prepare an edible manufacture in the event of a recall. The cannabis industry can look to future FDA regulation and the food industry's requirement for a recall plan. FDA requires the recall plan to include the description of steps to take and assign responsibility to (1) notify direct customers, (2) notify the public, (3) conduct effectiveness checks, and (4) execute disposition of the food. The cannabis industry is heavily regulated at the state level for much of this work already. Most food companies go beyond the FDA requirement and recognize the need for more information in the recall plan, so the company is as prepared as possible in the event of a recall. The food industry tests the recall plan with mock recalls which are of great value in learning weaknesses in the plan but not required.

The food industry could learn from the cannabis industry. For example, many food recalls are due to incorrect labeling. Good quality control in a cannabis company includes careful inspection of the product label of every package before the product leaves for the dispensary. In another example, I observed 100% inspection of vape cartridges at one facility before the cartridges went to packaging. I am not aware of any other quality management program in the manufacturing sector that does 100% inspection. This type of quality control in the cannabis industry contributes to a costly product, and the food industry cannot afford to build in this level of inspection. Once the cartridges were packaged, the QA department also inspected 100% of the labels on packages.

Supply chain program

Supply chain preventive control

Most parts of a supply chain program and procurement of ingredients and supplies are managed with a supply chain program. However, if the cannabis-infused edible manufacturer is counting on the ingredient supplier to control a hazard, a supply chain preventive control is applied. For example, in the making of chocolate bars, the chocolate supplier is expected to control for the biological hazard of *Salmonella*. The chocolate bar process at the edible manufacturer does not include a kill step, and if *Salmonella* were present in the chocolate, the edible would be adulterated and potentially cause illness. More information is in Chapter 7, Preventive Controls.

Allergen management program

Allergen preventive control

Most parts of an allergen management program are managed as GMPs. A facility that receives nuts as an ingredient must prevent cross-contact of the nut allergen with other ingredients starting at receiving, into storage, and at weigh up. The facility should verify allergen management at the supplier and have a supplier-approval GMP with associated SOPs. The allergen management GMP leads back to the supply chain program for management of allergens at the supplier. Starting at receiving, the facility implements GMPs for labeling and handling of the nuts. The facility should have a Receiving SOP to address allergen management. GMPs for allergen management are an example of "this is how we do things here." However, one or more steps are not managed as GMPs but instead noted as steps for allergen preventive controls. An example of an allergen preventive control is review of a label for the allergen declaration. More information is in Chapter 7, Preventive Controls.

GMP certification is facility-specific

GMP certification is a hot topic in the cannabis industry. I am not convinced that the cannabis industry knows what GMP certification is; they only know they want it. GMP certification is a certificate awarded to a facility where GMPs have been implemented, documentation is in order, and the facility has passed the related audit for certification. A GMP certificate is awarded to a stand-alone facility and not to employees at a facility. Multiple facilities within a company would individually earn a GMP certificate.

Customers drive certification. If a food company wants to get in with the big players like Costco, Walmart, Trader Joe's, and Whole Foods, they earn certification against one standard among a field of standards. Once we start talking about certification, we have left the path of FDA compliance. Being certified does not ensure compliance, and being FDA-compliant does not meet the standard for certification. An FDA inspector will not ask for a facility's GMP certificate. Document management is laborious, and each facility is left to figure it out on their own. When a company is developing documentation for FDA compliance and for certification, it is completely up to that company to manage the documentation for each. An FDA inspector does not care that a company is certified, nor does the FDA inspector review the copious documents required to earn and maintain certification.

GMP certification is earned

My cannabis clients want to implement GMPs and get GMP certified. Some edible manufacturers label their product as coming from a GMP compliant facility without earning certification. The food industry and dietary supplement industry would never think to label their product as coming from a GMP compliant facility. The food industry and dietary supplement industry know that product safety is built on GMPs and that GMPs are a given.

The article, *Quality Certification Programs for Dietary Supplements* [6], compared three GMP certification programs for dietary supplements in the United States from:

- ConsumerLab.com
- NSF International, and the
- US Pharmacopeial Convention (USP).

Earning and then keeping certification is a lot of work, hard work and usually completed by a certification specialist. Plan on the process to prepare for certification to take a year or more. If you are a small company or start-up and you do not have someone dedicated as their primary job to earn certification, it could take longer than a year. When you start to feel ready, do internal audits. A preaudit leading up to the official audit is worth the time and money. The lessons learned from the preaudit will get your systems in tip-top shape for the audit. Remember, quality assurance is a continual process and never done! Here are examples of standards not specific to cannabis but where cannabis companies may earn certification.

NSF/ANSI 173 for dietary supplements

The National Sanitation Foundation (NSF) and the American National Standards Institute (ANSI) developed a standard for the manufacture of dietary supplements. Companies can earn certification to NSF/ANSI 173 Dietary Supplements standard [7] which includes a GMP facility inspection [8].

ISO for food

ISO 22000 Food Safety Management is a globally recognized standard [9].

Global food safety initiative for food

Traditionally, food facilities have earned certification under a Global Food Safety Initiative (GFSI)-recognized certification program such as SQF, BRC, or FSSC 22000 [10]. BRC is more popular in Europe, while SQF is more popular in the United States.

Certification for growing operations

In my mind, I see growing and harvesting of cannabis the same as growing and harvesting of any other crop, e.g., corn, lettuce, strawberries, or apples. Traditionally, GMPs have not been applied at farming. There are farm programs for Good Agricultural Practices (GAPs), and this information can be found from resources of the United States Department of Agriculture (USDA), university extension agents, Produce Safety Alliance, and GlobalG.A.P. Under GFSI, GlobalG.A.P. is a standard for growing operations internationally. When marijuana is descheduled, university extension agents will be needed who specialize in the growing of cannabis. That will be a happy day! The lettuce growers of California and Arizona have the Leafy Green Marketing Agreement (LGMA) which provides training materials and audits farms toward certification. The LGMA brings farmers, USDA, and agricultural experts together to create best practices for lettuce growing and harvesting. In the future world where marijuana is descheduled, cannabis growing operations will benefit from a cannabis consortium that brings together best practices for the growing of cannabis.

Applying GMP certification to the cannabis industry

The million-dollar question remains: how do I earn certification for my cannabis company? The standards listed above are good options but not specific to cannabis products. Remember that certification is driven by customers, so certification should meet the needs of the customer. Here are a few certification options specific to cannabis.

Certification for medicinal products

The gold standard for the cannabis industry for GMP certification may be the European Union (EU) GMP guidelines [11]. Earning certification to the EU GMPs may be the most difficult path but also the most rewarding. Cannabis companies earning EU GMP certification may ship product into the EU which opens a market with enormous potential.

Certification for edibles

Back at home, the folks at TraceTrust have created a certification program specifically for cannabis-infused edibles with an emphasis on GMPs and dosing [12]. The NSF/ANSI 173 Dietary Supplements standard mentioned above can be applied to cannabis-infused edibles. SQF published a supplement to its GFSI-recognized standard which is specific to cannabis-infused edibles, *SQF Requirements and*

Guidance for Cannabis Manufacturing Supplement A compendium to the SQF Food Safety Code for Manufacturing, Edition 8.1 [13].

GMP certification does not ensure the safety of products

GMP certification shows a facility and its management are serious about product safety. That can only be a good thing to strive for excellent management of a facility and provide employee training so that GMPs are followed. Despite all the goodness of earning certification, the safety of a product is not ensured. This is a tough concept to convey, as the chapter title asserts that GMP compliance is not an option. GMPs are the foundation of product safety. It is an absolute requirement that a house be built on a strong foundation. Without a solid foundation, the integrity of the house will degrade. The foundation must be monitored and maintained. A crack in the foundation must be sealed. A neglected foundation over time will lead to a bigger problem. The building of the house can be done poorly or can be done exceptionally well. The quality and safety of the house are determined in the construction of the house on a strong foundation. The same is true for cannabis products. Product safety is built on a solid foundation of GMPs. Without implementing, monitoring, and maintaining GMPs, product safety will suffer.

The problem with certification is the auditing. Yes, all the preparation is useful and serves a purpose. Then the auditor arrives, and sees only a snapshot of the operation. The auditor does not have time to see everything or ask a question about every part under certification. The hope is that if your ducks are in a row for all the documentation reviewed and all the answers to the questions, everything else under certification is in order too. There are differences among auditors; there are differences among auditing bodies. It takes the right kind of person to be a good auditor, and it also takes the right kind of person to be the Quality Manager or SQF Practitioner in charge of the certification program.

Another way to explain the concept that GMP compliance does not ensure the safety of products is through some examples. My first example is a story told by a food safety colleague. My colleague was on-site at a food facility taking hundreds of swab samples. They were searching for the source of a pathogen contamination of the product. At the same time, the facility was being audited for certification. The facility received a perfect score from the audit! Auditing for certification is just a snapshot of operations and does not include testing for pathogens. The second example is the *Salmonella* outbreak and recall in 2008–09 traced to Peanut Corporation of America (PCA). PCA had recent to the outbreak been audited for certification and passed. Over 200 companies that used PCA's peanut butter as an ingredient recalled their products. The outbreak strain of *Salmonella* was confirmed in 714 patients across 46 states and at least one person in Canada. The infection was associated with nine deaths [14].

The Jensen Farms outbreak is my final example of how auditing for certification got it wrong. A 2011 outbreak of *Listeria monocytogenes* traced to cantaloupe sickened 147 people across 28 states with 33 deaths, one miscarriage, and another 10 people with listeriosis died with other causes noted as the reason for death [15]. The 22% mortality rate is typical for listeriosis.

A US Department of Justice news release from September 2013 stated:

> *In May of 2011 the Jensen brothers allegedly changed their cantaloupe cleaning system. The new system, built to clean potatoes, was installed, and was to include a catch pan to which a chlorine spray could be included to clean the fruit of bacteria. The chlorine spray, however, was never used. The defendants were aware that their cantaloupes could be contaminated with harmful bacteria if not sufficiently washed. The chlorine spray, if used, would have reduced the risk of microbial contamination of the fruit [16].*

Jensen Farms passed their audit in July 2011 with a score of 96% and a "superior" rating [17]. Each Jensen brother received 5 years of probation and 6 months of home detention, paid $150,000 in restitution and performed 100 hours of community service [18]. Ironically, the Jensen brothers were at the intersection of food safety and industrial hemp production in Colorado near the Kansas border in 2017. The same brothers who were guilty of selling adulterated cantaloupe were charged with drug distribution charges after 350 pounds of industrial hemp was shipped to California from Colorado with a stop in Kansas. The layover for the shipment in nearby Kansas made the shipment illegal, and they were charged with three felonies [19–21].

Focus on the immunocompromised consumer of cannabis-infused edibles

Writing makes me cement my beliefs and sharpen my understanding of a topic, so I am better prepared to present technical information in a webinar, at a conference or when consulting with clients. No topic was more confusing for me to write about in this book than GMPs, which made me dig in deeper. In the process of researching, discussing, and writing about GMPs, I have moved my own needle from "GMP certification isn't needed" as is the model in the food industry to "maybe GMP certification should be required" for the cannabis industry. I have also moved from cannabis-infused edibles being defined as food with added Δ^9- tetrahydrocannabinol and/or cannabidiol to cannabis-infused edibles being defined as dietary supplements with the FDA requirement for implementing an edible safety plan. As dietary supplements, edibles are consumed for potential medical benefits. Some people have severe, debilitating disorders which are mediated with cannabis to the point of the person living a functional life. The edible manufacturer does not know if the cannabis-infused edible will be consumed by a person already fighting a deadly disease and unable to fight an illness on top of the preexisting condition. Therefore, it is absolutely critical that cannabis-infused edibles are safe for consumption. That is, the edible must not cause illness from either the presence of a toxin or a pathogen. Every edible manufacturer has the responsibility to sell a safe product knowing the consumer may be immunocompromised (Box 4.3).

Box 4.3 Real People in Cannabis

Aaron G. Biros, Editor and Publisher of CannabisIndustryJournal.com

Time in the cannabis industry: I started Cannabis Industry Journal in 2015, so I've been working in the cannabis industry for just about 5 years now.

What drives me in my job: I began my career as an associate editor for FoodSafetyTech.com. I started Cannabis Industry Journal when the cannabis market was still a new idea. Only a handful of states had legalized the plant and I saw an immediate connection to food safety. With such a new market just starting to gain traction, there was a clear need to protect public health. By writing about how companies can help protect public health and safety, I found a niche that would eventually proliferate into the modern-day cannabis safety industry. Reporting on matters like standardization, safety, quality, and regulatory compliance, I wanted to help promote a well-informed and safe marketplace where patients and consumers could be protected from hazards like pathogens or other contaminants. Five years later, Cannabis Industry Journal is now a prominent resource for many in the cannabis space looking to improve the quality of their products and safeguard public health.

References

[1] Part 111 current good manufacturing practice in manufacturing, packaging, labeling, or holding operations for dietary supplements. https://www.accessdata.fda.gov/scripts/cdrh/cfdocs/cfCFR/CFRSearch.cfm?CFRPart=111. [Accessed 09.28.19].

[2] Part 117 current good manufacturing practice, hazard analysis, and risk-based preventive controls for human food. https://www.accessdata.fda.gov/scripts/cdrh/cfdocs/cfcfr/CFRSearch.cfm?CFRPart=117. [Accessed 09.28.19].

[3] Cannabis. https://www.inspection.gc.ca/plants/cannabis/eng/1531250234189/1531251514243. [Accessed 09.28.19].

[4] Bass G. FDA's perspective and experience on FSMA enforcement and inspections. International Association of Food Protection Annual Meeting, Louisville, Kentucky. 2019. July 22.
[5] Food and Drug Administration. Warning Letter, H & A Walters Fulfillment Center. August 15, 2019. https://www.fda.gov/inspections-compliance-enforcement-and-criminalinvestigations/warning-letters/h-walters-fulfillment-center-583217-08152019. [Accessed 09.13.19].
[6] Quality certification programs for dietary supplements. https://jandonline.org/article/S2212-2672(15)01642-1/fulltext. [Accessed 09.28.19].
[7] NSF/ANSI 173 - 2017 Dietary supplements. https://webstore.ansi.org/standards/nsf/nsfansi1732008addendum?gclid=Cj0KCQiA7aPyBRChARIsAJfWCgL vnZHo9E1X5dxFzrnfGN7GHpILqWDOZlNnsSfVb91o-tIkpYaa4hAaAj_4EALw_wcB. [Accessed 02.16.20].
[8] Dietary supplement safety. https://www.nsf.org/services/by-industry/nutritional-products/supplement-safety. [Accessed 02.16.20].
[9] ISO 22000 food safety management. https://www.iso.org/iso-22000-food-safety-management.html. [Accessed 02.16.20].
[10] Recognition: explore certification programmes. https://mygfsi.com/how-to-implement/recognition/. [Accessed 02.16.20].
[11] EudraLex - Volume 4 - Good manufacturing practice (GMP) guidelines. https://ec.europa.eu/health/documents/eudralex/vol-4_en. [Accessed 02.16.20].
[12] We make your edibles credible™. https://www.tracetrust.com/. [Accessed 02.16.20].
[13] SQF cannabis manufacturing supplement A. https://www.sqfi.com/wp-content/uploads/2020/01/SQF-Cannabis-Manufacturing-Supplement.pdf. [Accessed 02.16.20].
[14] Multistate outbreak of *Salmonella* typhimurium infections linked to peanut butter, 2008–2009 (final update). https://www.cdc.gov/salmonella/2009/peanut-butter-2008-2009.html. [Accessed 02.16.20].
[15] Multistate outbreak of listeriosis linked to whole cantaloupes from Jensen Farms, Colorado (final update). https://www.cdc.gov/listeria/outbreaks/cantaloupes-jensen-farms/index.html. [Accessed 09.28.19].
[16] Eric and Ryan Jensen charged with introducing tainted cantaloupe into interstate commerce. https://www.justice.gov/usao-co/pr/eric-and-ryan-jensen-charged-introducing-tainted-cantaloupe-interstate-commerce. [Accessed 09.28.19].
[17] Jensen brothers take responsibility but blame PrimusLabs. https://www.foodsafetynews.com/2013/10/jensen-brothers-sue-primus-over-third-party-audit-they-say-was-faulty/. [Accessed 09.28.19].
[18] Cantaloupe farmers get no prison time in *Listeria* outbreak that killed 33 people nationwide. https://www.usatoday.com/story/news/nation/2014/01/28/sentencing-of-colorado-cantaloupe-farmers/4958671/. [Accessed 09.28.19].
[19] Colorado brothers who sold deadly cantaloupe face Kansas hemp charges. https://kdvr.com/2019/07/27/colorado-brothers-who-sold-deadly-cantaloupe-face-kansas-hemp-charges/. [Accessed 09.28.19].
[20] Colorado brothers who sold deadly cantaloupe face Kansas hemp charges. https://www.cpr.org/2019/07/28/colorado-brothers-who-sold-deadly-cantaloupe-face-kansas-hemp-charges/. [Accessed 09.28.19].
[21] Colorado hemp producer fighting drug charges in Kansas after shipment stops in Liberal. https://www.hutchnews.com/news/20190724/colorado-hemp-producer-fighting-drug-charges-in-kansas-after-shipment-stops-in-liberal. [Accessed 09.28.19].

CHAPTER

Chemistry of cannabidiol and Δ^9-tetrahydrocannabinol

5

Concentrates are the products of extraction

Where I live in Wisconsin there is a regional food named *booyah*. I have lived throughout Wisconsin, northern Illinois, and into Minnesota. This is the only region that has booyah. It is such a big thing here that the new minor league baseball team in Green Bay is named the Booyah! As someone who did not grow up here, to me, booyah is chicken soup. If you say that booyah is chicken soup to a local, you immediately start an argument. Each family has their treasured booyah recipe; ingredients are guarded, and the cooking process makes it special. This is a bit of the attitude that is applied to a concentrate. To me, the final product of extraction is just a concentrate of cannabidiol (CBD) or a concentrate of CBD and Δ^9-tetrahydrocannabinol (THC). To the cannabis extractor, it is so much more than just a concentrate; there is science and art due to the variables of the process and how that process can get boutique concentrates with different CBD or THC isolates and different ratios of CBD to THC.

Concentrates take many forms from dried concentrate called *shatter* to free-flowing oils called *tinctures*. Cannabis-infused edible manufacturers create recipes with a concentrate as an ingredient. If FDA regulates the edible as a dietary supplement, the concentrate is analogous to a dietary ingredient. If the FDA regulates the edible as a food, the concentrate is analogous to a food additive. Both dietary ingredients and food additives have respective regulation already in place for each.

Biochemistry

Chemicals that are inhaled, injected, ingested, or absorbed into the body have a chemical reaction in the body, i.e., biochemistry. For some chemicals, such as glucose, there are biochemical pathways for the use of the energy and active parts of the chemical. Protein is digested only to be reassembled into muscle. In the case of cannabis, there is the well-known psychoactive property of THC as the THC reacts in the brain. Cannabis users seeking this biochemical reaction choose *Cannabis sativa* strains or hybrids. Cannabis users seeking relaxation choose *Cannabis indica* strains or hybrids. The popular term for the association of *C. indica* with the feeling of relaxation is *in da couch* as the cannabis user will relax and meld into the couch. Most of the physiologic

Food Safety Lessons for Cannabis-Infused Edibles. https://doi.org/10.1016/B978-0-12-819512-3.00005-5

properties of cannabis are known through experimentation. Individuals will vary in their reaction to the cannabinoids of THC and CBD in the same way that individuals vary with consumption of caffeine, sugar, alcohol, and food allergens. The human body is still a mystery in many ways.

Endocannabinoid system

There are people in the world who take a hard stance about never trying cannabis, never allowing the organic compounds of cannabis to enter their body. Little do they know that their body already has cannabinoids being made and circulating through their entire body. The human endocannabinoid system in the body naturally produces endocannabinoids that regulate appetite, pain, mood, and sleep [1] for homeostasis. We cannot live without endocannabinoids. The body has two types of cannabinoid receptors: CB1 and CB2. CB1 receptors are found in the central nervous system including the brain. CB2 receptors are on immune cells circulating through the bloodstream, and both CB1 and CB2 receptors are in the skin [2].

Phytocannabinoids

Trichomes are considered as biofactories of phytochemicals [3]. Phytocannabinoids are exogenous cannabinoids found in plant material. In addition to cannabis, phytocannabinoids are also found in black pepper, cacao, black truffles, kava, and the flowers Chinese rhododendron, helichrysum (a daisy), coneflower, electric daisy, and Japanese liverwort [4]. There is an argument for the love of chocolate based on chemistry! Over 500 distinct compounds have been found in cannabis, including compounds from the chemical groups of cannabinoids, terpenoids, flavonoids, and omega fatty acids. Over 100 different cannabinoids have been identified, and the chemical group of cannabinoids includes tetrahydrocannabinolic acid (THCA) and cannabidiolic acid (CBDA). THCA and CBDA are the acidic forms of THC and CBD, respectively, and the compounds that are naturally present in the plant material of hemp and cannabis. The process of decarboxylation is described later and is responsible for the chemical conversion of the acidic form to the biochemically active form of cannabinoids.

Terpenes

Terpenoids or terpenes create the unique aroma of the cannabis plant [5]. Terpenes create the flavor profile and aroma of cannabis products. Common terpenes naturally found in cannabis include [6]

- myrcene
- limonene
- pinene
- linalool
- caryophyllene
- humulene

Cannabis users choose cannabis strains for the desired effect and based on the predominance of terpenes [7].

Entourage effect

Cannabinoids and terpenes work synergistically in the physiologic effect of cannabis [6]. As will be discussed in more detail later, extraction captures compounds found in cannabis by the use of solvents. After extraction, further processing with winterization and distillation creates a concentrate that has some compounds removed and the desired compounds more concentrated and more isolated away from the not-so-desired compounds. The more isolated the concentrate becomes the more natural plant compounds are removed. There is a trade-off in removing compounds that work synergistically together for a greater effect by creating an isolate in high concentration and specific, known physiologic effect. There is a market for both the concentrate that contains close to the natural full spectrum of plant compounds and the cannabis isolate that is a high concentration of primarily one desired cannabinoid. Somewhere in between is the broad-spectrum concentrate, which is said to have THC removed while keeping the other cannabis plant compounds [8]. Patients using medical marijuana should be educated on the entourage effect when choosing a cannabis product that works to meet their needs.

Flavonoids

Beyond cannabinoids and terpenes are flavonoids. Like cannabinoids and terpenes, flavonoids are ring structures instead based on a 15-carbon phenolic structure [9]. Flavonoids are not identified as vitamins but play a similar role with antioxidative, anti-inflammatory, antimutagenic, and anticarcinogenic properties [9]. I first learned of flavonoids from my office mate at the University of Minnesota who was studying the chemistry of flavonoids in broccoli. That was in the 1980s and to this day, flavonoids are little discussed in food science but continue to be researched in chemistry and nutritional science [9]. Flavonoids are one of the reasons we are told to eat a variety of vegetables with a variety of color. In the same way that I eat broccoli for the nutrition, there are physiologic benefits from the consumption of flavonoids in cannabis. Look to this area of growing research in the future on flavonoids in cannabis such as cannflavin A and cannflavin B [10]. It is with descheduling of marijuana that cannabis research can proceed and the untapped potential of cannabis be realized.

Organic chemistry

The compounds of current interest in cannabis are organic compounds, i.e., compounds containing a carbon atom, that become biochemically active after inhalation, digestion, or absorption through the skin. The organic compounds of cannabis are complex molecules. The ring structure of CBD and THC makes them aromatic hydrocarbons. CBD and THC are only two compounds among a group of cannabinoids [11].

Extraction

Extraction is the process of separating compounds from the hemp or cannabis plant material into a concentrate that can be used for vaping, added to a topical, or added as an ingredient into an edible. Extraction is chemistry and a bit of physics. The extraction step starts with a large quantity of a raw agricultural commodity, i.e., the hemp or cannabis plant material, and ends with a very small amount of concentrate of the plant material worth thousands of dollars. Dried cannabis flower can go into extraction, if not being sold for smoking. Industrial hemp creates CBD concentrates with no more than 0.3% THC, where cannabis creates concentrates with CBD and THC over 0.3%. Both the concentration and the ratio of CBD to THC in the final concentrate are important. Creating a concentrate of CBD and THC is a feat of extraction, decarboxylation, and distillation, with endless variables.

Heat and pressure extraction

Extraction by heat and pressure is analogous to the process of squeezing olives to make extra-virgin olive oil. The advantage of this method is that the concentrate most closely represents the natural chemical contents of the plant, and patients looking for the entourage effect seek a concentrate made by this method.

Solvent extraction

Water

Extraction by water is analogous to coffee making; the plant material is left behind, and water-soluble compounds dissolve into the water. Cannabinoids in the natural acidic form will readily dissolve into water as the solvent. The advantage of water extraction is the safety of the solvent. The disadvantage is fat-soluble compounds will not be efficiently extracted, and the resulting concentrate will not be full spectrum.

Nonaqueous solvents

The list of chemical solvents that are used for hemp and cannabis extraction is extensive. Each solvent has its own advantages and disadvantages based on the solubility of cannabinoids, cost, codified usage, safety, and efficiency. Some solvents, e.g., ethanol, are food-grade and are best used to make concentrates for the purpose of edibles. Each state regulates the list of acceptable solvents and maximum levels of residual solvent in the concentrate. Local regulatory bodies determine solvent usage based on fire code, storage, and disposal. One disadvantage is that the concentrate may test over the limit for residual solvent and not be commercially viable.

Super critical carbon dioxide (CO_2)

Extraction by CO_2 is well researched in the food and other industries. For example, coffee beans are decaffeinated with CO_2 extraction. The disadvantage is the high cost of the equipment; an advantage is the high quality of the concentrate without residual solvent.

Following extraction, the process of winterization may be used on the concentrate to remove waxes and to further purify the concentrate; however, winterization is optional. The process of decarboxylation may follow extraction or may be done by heating the plant material before extraction. The following section is complementary to the information presented in Chapter 2, Overview of the Cannabis Industry.

Decarboxylation

Decarboxylation is the process of chemically removing a carboxyl group made of COOH from the natural acidic form of a cannabinoid to the neutral and biochemically active form of the cannabinoid (Fig. 5.1). Thus THCA becomes THC and CBDA becomes CBD. The high heat associated with smoking and vaping causes decarboxylation with the result of THC and CBD being inhaled. Once inhaled, the biochemically active form of the cannabinoid is absorbed into the blood vessels at the alveoli of the lungs, circulated through the body, and quickly makes its way to receptors in the brain and immune system [12]. Inhalation is the quickest way to experience the effects of cannabis. Decarboxylation is required of concentrates going into topicals or edibles. With topicals, the biochemically active form of the cannabinoid, CBD in most cases, is absorbed into blood vessels at the skin for a relatively quick local effect.

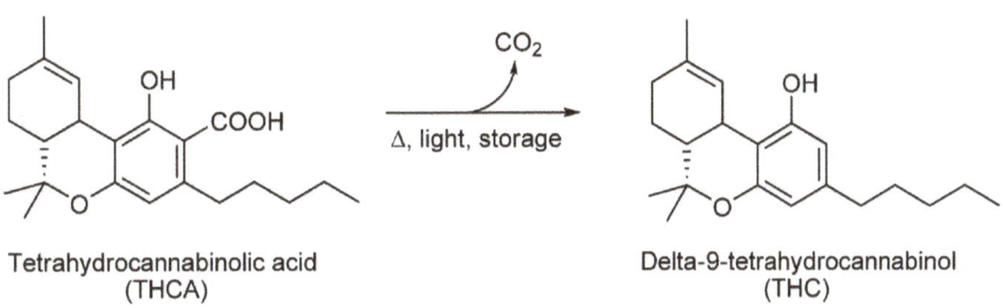

FIGURE 5.1

Decarboxylation of THCA to THC [13] with the release of carbon dioxide. The natural forms of the cannabinoids are found in the plant material as an acid and changed to the biochemically active forms by decarboxylation with heat, e.g., when smoking or vaping.

The temperature of the body is not high enough to cause the conversion of THCA to THC and CBDA to CBD, so the processing step of decarboxylation is required of the cannabis ingredient in an edible. In this way, the biochemically active form of THC and CBD is ingested. With the ingestion of an edible, the biochemically active forms of the cannabinoids will ride through digestion, will be absorbed into the bloodstream, and will make its way to CB1 and CB2 receptors in the brain and immune system. In addition, the liver plays a role in converting cannabinoids into more active compounds [12]. With digestion, absorption, and conversion, the effects of an edible are felt approximately 1 hour or longer after ingestion. Edibles remain an option for patients and consumers who wish to avoid smoking, but the time needed for the manifestation of the effects from the cannabis edible must be realized, and overconsumption must be avoided.

Distillation

After a concentrate is prepared with extraction and winterization, distillation may be used to further purify the concentrate into isolates. The process of distillation of a cannabis concentrate follows the standard steps of distillation, with the heating of the concentrate to create vapors, the vapors impact a cool condenser, and condensate drips off for collection. The distillation chemist knows what cannabinoids vaporize first through last and collects the condensate appropriately. The advantage of distillation is the desired cannabinoid has high purity and commands a high price. Cannabis manufacturers charge a high price for a small volume of isolate to recuperate the high cost of the distillation apparatus and high salary of the chemist.

Analytical chemistry

Chemists are needed in the cannabis industry for extraction operations and laboratory analysis. The processes of extraction, winterization, and distillation follow the laws of chemistry. Cannabis companies hire organic chemists to operate and manage extraction operations. Cannabis laboratories hire analytical chemists to test samples and manage chemistry departments. The role of the cannabis laboratory is to be a third-party, independent testing laboratory for the cannabis industry. Cannabinoids are measured with the analytical chemistry methods of liquid chromatography, gas chromatography, and mass spectroscopy. There is an investment by the cannabis laboratory in the order of hundreds of thousands of dollars per instrument and in the hiring of highly paid chemists to perform the analyses.

Analytical chemistry is used to test samples for the hazards of pesticides, heavy metals, residual solvents, and mycotoxins. The extraction process will concentrate these chemical hazards, along with cannabinoids, depending on the extraction method. Before extraction, some companies will test incoming plant material to understand the concentrations of both chemical hazards and THC and CBD. Potency testing for THC and CBD uses analytical chemistry and is mandated on the finished product of dried flower, concentrates, topicals, and edibles per the state code.

References

[1] Manzanare J, Julian MD, Carrascosa A. Role of the cannabinoid system in pain control and therapeutic implications for the management of acute and chronic pain episodes. Current Neuropharmacology 2006;4:239–57.

[2] Absorption of cannabinoids through the skin is an effective means for delivery into the body. https://weedmaps.com/learn/dictionary/cannabinoid/. [Accessed 09.15.19].

[3] The name of cannabis: a short guide for nonbotanists. https://www.ncbi.nlm.nih.gov/pmc/articles/PMC5531363/. [Accessed 09.15.19].

[4] Plants other than cannabis that produce cannabinoids. https://www.royalqueenseeds.com/blog-plants-other-than-cannabis-that-produce-cannabinoids-n714. [Accessed 09.28.19].

[5] Cannabis and its compounds. https://www.uclahealth.org/cannabis/cannabis-and-its-compounds. [Accessed 09.15.19].

[6] The difference between cannabinoids and terpenes. https://www.analyticalcannabis.com/articles/the-difference-between-cannabinoids-and-terpenes-311502. [Accessed 10.05.19].

[7] Marijuana terpenes and their effects. https://www.alchimiaweb.com/blogen/marijuana-terpenes-effects/. [Accessed 10.05.19].

[8] Full spectrum vs broad spectrum vs CBD isolate: The difference explained. https://medium.com/cbd-origin/full-spectrum-vs-broad-spectrum-vs-cbd-isolate-the-difference-explained-a09e415bf235. [Accessed 10.05.19].

[9] Panche AN, Diwan AD, Chandra SR. Flavonoids: an overview. Journal of Nutrition Sciences 2016;5:e47. https://doi.org/10.1017/jns.2016.41.

[10] Understanding cannflavin biosynthesis unlocks pain relief potential of cannabis. https://www.technologynetworks.com/drug-discovery/news/understanding-cannflavin-biosynthesis-unlocks-pain-relief-potential-of-cannabis-322057. [Accessed 10.05.19].

[11] Cannabinoids 101: what makes cannabis medicine? https://www.leafly.com/news/cannabis-101/cannabinoids-101-what-makes-cannabis-medicine. [Accessed 10.05.19].

[12] Metabolism of THC. https://sapiensoup.com/images/thc/Infographic-Human-Metabolism-THC-1300.png. [Accessed 09.15.19].

[13] Hazekamp A, Fischedick JT, Diez ML, Lubbe A, Ruhaak RL. Comprehensive Natural Products II, Chemistry and Biology, Volume 3, p. 1045. Philadelphia, PA: Elsevier; 2010 [3.24-Chemistry of Cannabis].

CHAPTER

The hazard analysis

6

Beyond Hazard Analysis and Critical Control Points (HACCP)

Hazard Analysis and Critical Control Points (HACCP) was a food safety program applied to food production for the manned space missions to the moon and thereafter throughout the food industry starting in the 1960s. Except for the FDA rules requiring seafood HACCP and juice HACCP and the US Department of Agriculture rule for meat HACCP, the food industry was under the voluntary use of HACCP. As a food safety program, HACCP addressed the potential hazards present in a food facility and controlled through processing steps. Typical control under HACCP included the kill steps of pasteurization, cooking, baking, and roasting.

Many states require the cannabis industry to follow HACCP for cannabis-infused edibles. HACCP is a good place to start in the control of biological, chemical, and physical hazards, and many state inspectors may be comfortable with the enforcement of HACCP due to its relatively long history in the food industry. I see HACCP as a stepping-stone for the cannabis industry as a means to grow into the broader federal requirement of food manufacturers. In 2017, the author explained the concepts of HACCP in a four-part series of articles published on the *Cannabis Industry Journal* site and entitled *Hazard Analysis and Critical Control Points (HACCP) for the Cannabis Industry* [1–4].

In 2011, President Obama signed the Food Safety Modernization Act (FSMA), and in 2015 the FDA published 21 CFR 117 *Current Good Manufacturing Practice, Hazard Analysis, and Risk-based Preventive Controls for Human Food* [5] (PCHF), one of the seven foundational rules of FSMA. The PCHF rule requires the implementation of a written food safety plan to include process preventive controls as does an HACCP plan, but goes beyond HACCP. In addition to process preventive controls, a food safety team must evaluate the control of hazards by (1) allergen, (2) sanitation, and (3) supply chain preventive controls. At least one member of the food safety team must be qualified as a Preventive Controls Qualified Individual (PCQI). The PCQI has the knowledge, training, and experience to identify hazards and implement preventive controls. The food safety team evaluates potential hazards by going one step back to suppliers and one step forward to customers in the control of food safety. In this way, the control of hazards goes beyond HACCP to include the safety of ingredients, control of environmental pathogens, control of allergens, and

Food Safety Lessons for Cannabis-Infused Edibles. https://doi.org/10.1016/B978-0-12-819512-3.00006-7

establishing a recall plan. The first requirement for a food safety plan or edible safety plan is to conduct a hazard analysis.

Reminder

As noted in Chapter 4, Good Manufacturing Practice Compliance is Not Optional, a *food safety team* is analogous to an *edible safety team* and a *food safety plan* is analogous to an *edible safety plan*. Food safety information, knowledge, and experience are presented here for application to the cannabis-infused edibles industry.

Definition of hazard analysis

The process of collecting and evaluating information on hazards and conditions leading to their presence to decide which are significant for food safety and therefore must be addressed in the Food Safety Plan [6].

The hazard analysis starts with the work of controlling hazards in the product. The hazard analysis is conducted by the food safety team or edible safety team at a facility and typically documented on a hazard analysis form. There is no mandated form; a typical form includes the information presented in six columns but may also include additional information (Fig. 6.1).

The letters B, C, and P in the form in Fig. 6.1 are defined as follows:

Biological (B) hazards, including viruses, bacterial pathogens of concern in ingredients, environmental pathogens of concern with ready-to-eat (RTE) products, protozoans, and parasites.

Chemical (C) hazards, including allergens, radioactive chemicals, mycotoxins, pesticides, drug residues, industrial chemicals, decomposition by-products, and unapproved food or color additives.

Physical (P) hazards, including choking hazards and potentially harmful pieces of metal, glass, hard plastic, field debris, or nut shells.

Hazard Analysis

(1) Ingredient / Processing Step	(2) Identify potential food safety hazards introduced, controlled or enhanced at this step		(3) Do any potential food safety hazards require a preventive control?		(4) Justify your decision for column 3	(5) What preventive control measure(s) can be applied to significantly minimize or prevent the food safety hazard? *Process including CCPs, Allergen, Sanitation, Supply-chain, other preventive control*	(6) Is the preventive control applied at this step?	
			Yes	No			Yes	No
	B							
	C							
	P							

FIGURE 6.1

One example of a hazard analysis form; a specific form is not mandated [6]. *CCPs*, critical control points.

The following is an excerpt of an FDA warning letter citing violations resulting from an inadequate hazard analysis [7].

1. *Your hazard analysis did not identify a known or reasonably foreseeable hazard for each type of food manufactured, processed, packed, or held at your facility to determine whether there are any hazards requiring a preventive control, as required by 21 CFR 117.130(a)(1). The hazard evaluation must include an evaluation of environmental pathogens whenever a ready-to-eat (RTE) food is exposed to the environment, as required by 21 CFR 117.130(c)(1)(ii).*

The warning letter goes on to identify *Salmonella* as the environmental pathogen of concern. The next violations were a fall out from the first violation [7].

2. *Your firm did not identify and implement a preventive control for a hazard when one was needed for your RTE products, as required by 21 CFR 117.135(a)(1).*
3. *You did not establish and implement written corrective action procedures to be taken for environmental monitoring verification activities, to comply with 21 CFR 17.165(a)(3) and (b)(3)(vii) and 21 CFR 117.150(a)(1),*
   and the company did *not include documentation as to how you will ensure appropriate corrective actions are taken in the future.*

The following is an excerpt of a different FDA warning letter citing violations resulting again from an inadequate hazard analysis [8].

During our inspection of your facility, FDA investigators found serious violations of the Current Good Manufacturing Practice, Hazard Analysis, and Risk-Based Preventive Controls for Human Food regulation (CGMP & PC rule), Title 21, Code of Federal Regulations, Part 117 (21 CFR Part 117).

Based on FDA's inspectional findings, we have determined that the RTE products manufactured in your facility are adulterated within the meaning of Section 402(a)(4) of the Federal Food, Drug, and Cosmetic Act (the Act) [21 U.S.C. § 342(a)(4)] because they were prepared, packed, or held under insanitary conditions whereby they may have become contaminated with filth or whereby they may have been rendered injurious to health.

From this warning letter, we see similar language on the hazard analysis violation as we saw in the previous example.

You did not prepare, or have prepared, and did not implement a food safety plan for the food manufactured in your facility, such as RTE salads, as required by 21 CFR § 117.126. A food safety plan must include the following:

(1) The written hazard analysis, as required by § 117.130(a)(2);
(2) The written preventive controls, as required by § 117.135(b);
(3) The written supply-chain program, as required by subpart G;
(4) The written recall plan, as required by § 117.139(a);
(5) The written procedures for monitoring the implementation of the preventive controls, as required by § 117.145(a);
(6) The written corrective action procedures, as required by § 117.150(a)(1); and
(7) The written verification procedures, as required by § 117.165(b)

Importance of doing a good job the first time

The hazard analysis is the BIG DEAL in food and edible safety and the basis for management decisions in the control of hazards. Expect the state or FDA inspector to read every word in the hazard analysis. The PCQI and team members must be ready to justify the presence or absence of a hazard for every ingredient and at every process step in the hazard analysis, the control of a hazard with a Good Manufacturing Practice (GMP) or preventive control, and the type of preventive control applied. Preventive controls are the procedures implemented at a food facility to control biological, chemical, and physical hazards. Every process preventive control must be validated. After the hazard analysis is finished, every process, allergen, and sanitation preventive control will have standard operating procedures (SOPs), records, monitoring, verification, and correction or corrective action. Supply chain preventive controls for ingredients require evidence of food safety through the collection of documentation. Elaboration on this theme is found in Chapter 7, Preventive Controls.

An underlying importance of doing a good job on the hazard analysis is that the team will identify action items to be presented to the upper management. The team may identify capital expenses needed for food safety improvement. The team may decide to change ingredient suppliers or formulations. I recommend that food safety or edible safety teams create an action plan with a list to present to the appropriate department managers and upper management for improvement projects. As the team works through the hazard analysis, areas of improvement can be noted and presented to the upper management for management decisions. The action plan will serve as a tool for continuous improvement in food safety.

Importance of teamwork

The right people need to be on the food safety team or edible safety team. The team requires at least one PCQI, and it is wise for a company to have additional PCQIs as a backup to the PCQI. I delivered the PCQI training course on-site at a food facility where 17 PCQIs were trained. It would be a disaster to have 17 PCQIs on the team because a team with too many people would get bogged down in too much discussion and not enough results. I recommend eight or less people be assigned to a team. In addition to the PCQI, who is probably a quality manager, can the team include maintenance, sanitation, operations, purchasing, and receiving personnel? Think outside the box when creating the team, so that a variety of perspectives across the facility is included.

If technical expertise is needed, a consultant may be hired as a member of the food safety team or edible safety team. I worked with an outstanding team at a food facility for the writing of their hazard analysis for an RTE product with only RTE ingredients. There was no kill step on-site at the client's facility for the ingredients or product. With no kill step, the client relied solely on the control of biological hazards at the ingredient suppliers, through employee hygiene, and with the sanitation of food contact surfaces and the processing environment. My expertise was well applied for biological hazards. We thought we were prepared for a conference call with the customer's audit team, and I participated as the consultant and a team member. Guess what completely blindsided us? Allergen control—not at my client's facility but at the facilities of the suppliers. The customer wanted to know what documentation we had for confidence of allergen control at the suppliers. That was a hard hit at the time for me as a consultant. I have since grown my knowledge of allergens and of FDA's growing interest in the control of cross-contact. Now I know to include undeclared allergens as a hazard, requiring a supply chain preventive control.

Corporate technical expertise can be shared, and there may be a member on the team from corporate. Here are words of caution about corporate involvement in writing a hazard analysis. It is tempting for corporate to roll out a company-wide hazard analysis and edible safety plan. The first reason this is wrong is that edible safety plans are facility-specific, so it is not possible to have one-size-fits-all because of the variables at each facility. There are enough variables from facility to facility even if ingredients come from the same suppliers, the same equipment is used, and the same products are made. The traffic patterns and food safety culture of each facility are different, just to start. The second reason is that the process of conducting a hazard analysis is a learning opportunity for the team members. Team members will learn about practices and procedures of which they had no idea were occurring and affect edible safety, which provides opportunities for improvement. Team members need to know the hazard analysis intimately because each member will be on the hot seat to answer questions and provide justifications to an inspector (Box 6.1).

The hazard analysis is best to be completed as a team and not to divide and conquer the work; each team member is there to contribute his/her expertise from his/her department. Preparing the hazard analysis is a lot of mental work and physical time spent in team meetings. Scheduling issues to get everyone on the team together with consideration to shift work is a reason to keep the team lean and not have too many team members. Expect the hazard analysis to be grueling teamwork with dedicated time for the team to meet. If the analysis is too easy, the hazard analysis lacks rigor and a hazard may have been missed and left uncontrolled.

Documents and records can be retrieved by individual team members and provided to all. For example, one person can generate the list of ingredients received and another person can create an updated flow diagram of the process steps. These are first steps to get the ball rolling and get documentation in front of the team for analysis. The entire team should use the updated flow diagram to walk through the facility and follow the flow of processing as a team. The flow diagram is not required in a food safety plan, but most facilities have this useful tool. After the walk-through, the team should gather to note observations and changes to the flow diagram. After the flow diagram reflects the true flow of operations at the facility, the team will evaluate potential hazards for each process step, one by one as a team, to complete the hazard analysis one process step at a time.

Box 6.1

A *flow diagram* is a series of connected boxes showing the steps at a facility from receiving to shipping. The flow diagram is not required by law but is very helpful as a communication tool for the team in conducting a hazard analysis. An example of a flow diagram is Fig. 2.1.

Once the team has the list of ingredients, one ingredient at a time will be analyzed completely to identify hazards, going one by one through the list of ingredients. Grouping of like ingredients with the same hazards is allowed. For example, a bakery might group different types of flours received from the same supplier or a chocolate maker might group flavoring extracts. The team will meet for hours over many days, so schedule chunks of time until the hazard analysis of all ingredients and all process steps for one food safety plan is finished. If a facility has more than one food safety plan, it is recommended to finish one plan at a time entirely and then go on to the next plan.

Resources for the hazard analysis

Personnel

Having the right people on the food safety team or edible safety team as discussed earlier is just the beginning. The team will need to gather information held by other employees, such as the list of suppliers of ingredients, customer complaint records, laboratory results, recall notices and warning letters, and technical literature. The best resource is the employees. The team will seek out the appropriate employees from which to collect observations, documents, and data.

Food and Drug Administration rule and guidance

The PCQI will know how to access the FDA PCHF rule [5] and its guidance documents from the PCQI training materials and FDA FSMA PCHF rule website. Have these documents ready to be accessed during team meetings.

I recommend a food safety team or edible safety team use the FDA Appendix 1 [9] to identify hazards in ingredients. The full title of this document is *Hazard Analysis and Risk-Based Preventive Controls for Human Food: Draft Guidance for Industry, Appendix 1: Potential Hazards for Foods and Processes* (Appendix 1). This is a critical resource because the state or FDA inspector will expect every hazard in Appendix 1 for the ingredient to be listed as a hazard, addressed as a GMP or preventive control, and justified. Delegate to the person on the team who is good at searching PDF documents the role of Appendix 1 searching during team meetings. Appendix 1 is 233 pages and tricky to figure out how to use. There are three sections of tables.

1. Biological Hazards Tables are all table 1, the first section.
2. Chemical Hazards Tables are all table 2, the second section.
3. Process Hazards Tables are all table 3, the third of three sections.

The order of each of the three sections repeats by commodity. The lettered order from A to Q covers the following commodities:

A. Bakery Items
B. Beverage Items
C. Food Additives

D. Chocolate and Candy
E. Dairy
F. Dressings, Condiments and Dips
G. Egg and Egg Products
H. Fruits and Vegetables
I. Game Meat Products
J. Grains, Beans and Grain Products
K. Multi-Component Foods
L. Nuts and Seeds
M. Oil and Oil Products
N. Snack Foods
O. Soups and Sauces
P. Spices and Herbs
Q. Sweeteners (nutritive and nonnutritive)

It is the combination of the letter and number that gets you to the right table, e.g., for the first commodity of Bakery Items, you always go to the A table:

> *Table 1A: Information that you should consider for potential ingredient or other food-related biological hazards for Bakery Items.*
> *Table 2A: Information that you should consider for potential ingredient or other food-related chemical hazards for Bakery Items.*
> *Table 3A: Information that you should consider for potential process-related biological, chemical, and physical hazards for Bakery Items* [9].

If you are purchasing the food as an ingredient then include the biological hazards from the table 1 and chemical hazards from the table 2 for that one item in your hazard analysis of the ingredient. The control of the hazard will be at the supplier's facility with a supply chain preventive control or at your facility. In rare cases, the control of the biological hazard will be with the customer, e.g., in the case of frozen, unbaked bakery items for which the customer or consumer bakes the product. The FDA has not provided a table in Appendix 1 for physical hazards in purchased ingredients directly. If you are purchasing the ingredient, you can expect your supplier to control the physical hazard. If you are the manufacturer of an ingredient or food product then include the process-related hazards from the table 3 in your hazard analysis of your processing steps at your facility. The three tables are for your processing steps and control of hazards at your facility.

I learned a lesson in the fact that I do not know it all when I wrote the hazard analysis for a cannabis-infused chocolate bar (Appendix 2) for my work with ImEpik, an online training platform. After training 300+ PCQIs, I was confident in my first draft of the hazard analysis, and then I checked the FDA Appendix 1 tables of hazards in the ingredients. It was nice to confirm my knowledge that FDA identifies no hazards for corn syrup, but I learned of hazards considered by FDA for other ingredients. I had to add those additional hazards to ingredients in my hazard analysis and address the hazards. Your team will go through a similar process of knowing hazards associated with the ingredients you purchase, and you will include in your hazard analysis any hazards you learn from your resources, including Appendix 1 [9].

State code for marijuana regulations

There is one ingredient you will not find in the FDA Appendix 1 and that is cannabis concentrate. With the hazard analysis of cannabis concentrate destined for edibles, the team will identify biological, chemical, and physical hazards. The state code is the resource for required testing of potential hazards in cannabis concentrate. As a scientist and lead instructor for PCQI training, I can tell you that there is a big difference between identifying a hazard that by definition causes illness or injury, including death, and identifying hazards requiring testing per the state code. Use caution. Just because state code requires testing does not automatically mean the hazard is there in the purchased cannabis concentrate. For example, state code requires testing for pesticides. What if it was possible to source cannabis that had never been exposed to pesticides by application, air, soil, or water? Documentation of no pesticide exposure could substitute for pesticide testing. On the other hand and as the industry grows, hazards may be identified without required testing by state code. For example, some cannabis companies are testing cannabis concentrate destined for vaping cartridges for vitamin E oil, a harmful ingredient when heated and inhaled, even though the state code does not mandate the testing of vitamin E. The team should identify hazards in the hazard analysis by risk analysis based on severity and likelihood [10] (Fig. 6.2). There is no required risk matrix, but the risk matrix is a useful tool for the team in answering the question in column 3 on the hazard analysis form (Fig. 6.1).

In the food industry, testing of a lot of finished product or of an environmental sample is often required by customers, not by regulation. The bigger the retail customer, the more demanding they can be. In the cannabis industry, the state code dictates testing of flower, concentrate, and edible product. The testing of flower seems reasonable for the product it will go into. Absent peer-reviewed research of hazards in concentrate, I completely agree with the state code for testing of concentrate. The state code for concentrate will vary from state to state, but for where the cannabis industry stands at this time, the requirements of each state tend to be reasonable. As a supply chain preventive control, the concentrate, butter, or oil must be tested for safety and potency. Potency is the concentration of Δ^9-tetrahydrocannabinol (THC) and cannabidiol (CBD) in the concentrate and will be critical to batching of the edible.

That leaves the testing of the edible. Think of the state code as a customer of the edible. Some of the testing will verify the control of a hazard. When *Salmonella* or Shiga toxin-producing *E. coli* (STEC) are tested from an edible, there is verification of the control of pathogens. When mycotoxins, pesticides, or heavy metals are tested from an edible, there is verification of the control of a chemical hazard. Unfortunately, the process is going backward, and that is just wrong. With the state code dictating the testing, the state is identifying the hazard requiring control. You may think that the state code is no different than FDA publishing Appendix 1 discussed earlier [9]. The difference is that FDA does not dictate any testing, expects industry to build in safety instead of conducting finished product testing, and requires industry, specifically a food safety team, to identify the hazards. FDA Appendix 1 is a resource to help industry identify hazards.

RISK RATING = LIKELIHOOD X SEVERITY

Severity					
Catastrophic 5	5	10	15	20	25
Significant 4	4	8	12	16	20
Moderate 3	3	6	9	12	15
Low 2	2	4	6	8	10
Negligible 1	1	2	3	4	5
©2019 Property of ImEpik, LLC	Improbable 1	Remote 2	Occasional 3	Probable 4	Frequent 5

Likelihood

Risk	Definition	Code	Action Required
Catastrophic	Resulting in death, illness or injury		Requires a preventive control
Unacceptable	Resulting in illness or injury		Requires a preventive control
Undesirable	Potential unacceptable risk		Requires food safety team decision for hazard control by preventive control or GMP
Acceptable	Risk controlled by GMPs under normal operating conditions		Controlled by GMPs
Desirable	Risk controlled by GMPs under normal operating conditions		Documented in the hazard analysis as not requiring a preventive control or GMP

FIGURE 6.2

The risk matrix uses severity and likelihood to decide if a hazard is controlled by Good Manufacturing Practices (GMPs) or preventive controls.

Used with permission from ImEpik LLC.

When I wrote the hazard analysis for the cannabis-infused chocolate bar (Appendix 2), I had to pick a state code. I picked Colorado marijuana regulations, only because I possess a bound copy of the 2019 Colorado Marijuana Laws and Regulations [11]. I had a general idea of the testing requirements, but I was shocked at the detail and abundance of testing required by the state code. I have great respect for the work of cannabis laboratories. The hazard analysis for the cannabis-infused chocolate bar is written with all the Colorado state code parameters as hazards requiring a preventive control. That makes me feel uneasy because my gut feeling is that many of the hazards required by the state code to be tested are not hazards requiring a preventive control! Only with research and data will an edible manufacturer be able to separate out cannabis hazards controlled by GMPs from hazards controlled by preventive controls. For example, in the food industry, pesticides are noted as a hazard on fresh produce. When pesticides are appropriately applied, the hazard is controlled. Most food facilities can control the hazard of pesticides with a GMP in the form of a pesticide management program. It is not that pesticides are disregarded; it is that the supplier has a history of compliance with a GMP and the risk of the hazard is too low to require a preventive control. The cannabis industry will get there too.

Finished product testing is a burden on the cannabis industry. With federal descheduling of marijuana, there will be federal oversight instead of individual state codes. With federal oversight, we can look forward to FDA applying the same standard to the cannabis industry as it has for the industries of dietary supplements and food. I expect FDA to put the burden of hazard analysis on the edible industry in the same way that the burden of hazard analysis is on the food industry. One definition that is unimaginable at this time is the definition for a very small edible business. Because of public health considerations, the FDA defines a very small (food) business as having less than $1 million in annual revenue. This very small business must follow GMPs and any local or state code but is exempt from a written hazard analysis and implementation of any preventive controls. Following descheduling of marijuana expect FDA to define very small and small cannabis-infused edible businesses by revenue. If a chocolate bar costs $20, a very small business is defined as manufacturing 50,000 units annually or less than 1,000 bars per week. The current state code may be harsher than the federal regulation, and unfairly hurt the start-up companies and what the food industry calls the mom-and-pop shops (Box 6.2).

Box 6.2 Important points about edible safety plans

The edible safety plan is specific to a facility because of the unique architecture of the building, the layout of equipment, suppliers, and culture. Within a facility, the edible safety team has the choice of writing edible safety plans for individual products or groups of products. For example, bakery items using the same equipment can be one group of products. In most cases, the same will be true for the grouping of chocolates, gummies, hard candy, or beverages. One group of products has one hazard analysis paired to one edible safety plan. The team PCQI can lead the edible safety team through the process of grouping products based on shared equipment and hazards.

As a lead instructor for the training of PCQIs, I had many courses under my belt and had trained dozens of PCQIs when I started to help individuals in the food industry write their hazard analysis. Many of the clients I work with are small businesses doing their best to meet FDA compliance with limited resources and time stretched too thin. As a member of the Expert Services team of ConnectFood,

I am hired on a project basis to work with someone from the company, e.g., owner or quality manager. I always start with a blueprint picture of the facility and flow diagram of the process from receiving through shipping. In working with the company employee, I identify all the processing steps; I visualize the process flow in the facility. The next task is to identify all the ingredients for the products in the food safety plan. Some of my clients do a hazard analysis of the ingredients in a separate document from the hazard analysis of the process, and some clients have the process and ingredients together in one hazard analysis. The choice is yours.

I have learned a great deal over the years about writing a hazard analysis, and I continue to learn. For example, for three years, I was critical of the FSPCA including "Other" as a category for preventive control; see column 5 (Fig. 6.1). I told my workshops that "Other" could be used for when a food safety team cannot decide on one category to name the preventive control or gives the FDA wiggle room to add another type of preventive control down the road. Those were two reasons I made up to explain "Other," until I had occasion to use "Other." I was working as a project manager for ImEpik to update their online course for PCQI. We built in scenarios to apply the learning, and one of the scenarios was for frozen, unbaked pastries, a bakery product. I identified that the biological hazard was not controlled by the supplier, so the preventive control was not a supply chain preventive control. There was no process preventive control at the facility because the pastries were unbaked. Even though the pastries would be baked by the consumer, I did not want to label the control as a process preventive control. The biological hazard was not controlled by an allergen or sanitation preventive control. That left the option of "Other." I liked that! In other words, I labeled control at the customer or consumer as Other preventive control.

The writing of the hazard analysis in principle comes after the writing and implementing of GMPs because GMPs are the foundation of food safety upon which all other food safety programs are built. In addition, the resulting preventive controls from the hazard analysis go hand in hand with the facility's GMPs. All food safety issues will be addressed with either a GMP or a preventive control. The other reason the hazard analysis is written after the GMPs only relates to our place in history. The food industry has only been required to write a hazard analysis since 2016 for businesses with more than 500 employees, since 2017 for businesses with greater than $1 million in annual revenue, and since 2018 for fluid dairy plants because of their regulation under the Pasteurized Milk Ordinance and state code. GMPs have been mandated long before the written hazard analysis, so the food industry has built the hazard analysis on top of functional and operational GMPs. In fact, writing the hazard analysis was difficult for some companies because the company had been doing a great job of controlling hazards with effective GMPs, such as cleaning and sanitizing of equipment. Functional GMPs will never go away; they just may be renamed as preventive controls.

The three-year roll out of the FDA PCHF rule was designed to bring the largest companies into compliance first with the understanding that they would have the resources to get everything done by the 2016 deadline. Then small businesses came on board, followed by fluid dairy plants. Alongside industry, the FDA inspectors have been coming up to speed as well, starting with the same training delivered to industry for PCQIs, followed by more FDA training, shadowing of fellow FDA inspectors during inspections, and having a cadre of FDA experts to call with questions during an inspection. In addition to the training of more FDA inspectors each year to perform preventive controls inspections, state inspectors are being trained to perform the FDA preventive controls inspection in states where there is an agreement between the state and FDA. In the first fiscal year of enforcement, the FDA goal for preventive controls inspections in fiscal year 2016/17 was 200 domestic inspections and 100 foreign inspections. As was reported in July 2019, the FDA inspections for fiscal year 2018/19 were 423 domestic and 137 foreign preventive controls inspections, with the respective goals of 500 and 150 inspections [12]. Industry should expect the goal to increase annually with the training of more FDA and state inspectors. FDA is doing a factor of seven times more GMP inspections, which do not include a review of the hazard analysis or swabbing to detect pathogens [12].

The steps of writing a hazard analysis

Using resources and as a team follow these steps for writing a hazard analysis (Fig. 6.1).

1. List the process steps and ingredients in Column 1.
2. Identify hazards in Column 2 by evaluating every process step and every ingredient for biological, chemical, or physical hazards.
3. Conduct a risk assessment of every hazard in Column 2 to identify hazards requiring a preventive control and mark those in the Yes column.
4. Write detailed justifications for both No and Yes marks for each identified hazard. In some cases, the team may want to write a detailed justification for not identifying a hazard in Column 2. Note when GMPs are controlling a hazard.
5. Choose and note the type of preventive control for hazards with Yes in Column 3. Avoid the use of "Other" except for the rare control of a hazard by the customer.
   a. Process
   b. Allergen
   c. Sanitation
   d. Supply chain
   e. Other
6. Mark No if the hazard will be controlled at a later step. Mark Yes if the hazard is controlled at the step. Supply chain preventive controls are applied at receiving. Sanitation preventive controls for environmental pathogens are applied after products are packaged and removed from the processing environment but at the process step where there is potential for cross-contamination.

Follow up to a good hazard analysis will include gathering documents and data on a kill step validation, parameters, and values; allergen management; ingredient specifications and laboratory results; and the cleaning and sanitizing parameters and values for food contact surfaces and processing and packaging rooms. The team may capture these documents and data while conducting the hazard analysis and in preparation of writing the preventive controls for the control of identified hazards.

Summary

When an edible safety team writes a thorough and complete hazard analysis, there is identification of (1) hazards requiring a preventive control, (2) the type of preventive control, and (3) the step where the hazard will be controlled. Development of the hazard analysis and assignment of preventive controls are further discussed in Chapter 7, Preventive Controls. Control of environmental pathogens in RTE food and edibles is discussed in Chapter 8, Nobody is Talking About Environmental Monitoring (Box 6.3).

Box 6.3 Real people in cannabis

Irmak Erdem, Research technician, MJardin group, colorado

Time in the cannabis industry: April 2018, my first *Cannabis* job at MJardin Group.

What drives me in my job: My background in biotechnology (B.S.) and plant molecular biology (M.S.) has inspired me to pursue the unknown in the *Cannabis* industry. I see problem-solving to be both fun and challenging, requiring logic, creativity, and patience. I am driven by data, organization, and growth.

References

[1] Hazard analysis and critical control points (HACCP) for the cannabis industry: Part 1. https://cannabisindustryjournal.com/feature_article/hazard-analysis-and-critical-control-points-haccp-for-the-cannabis-industry-part-1/. [Accessed 10.05.19].

[2] Hazard analysis and critical control points (HACCP) for the cannabis industry: Part 2. https://cannabisindustryjournal.com/feature_article/hazard-analysis-and-critical-control-points-haccp-for-the-cannabis-industry-part-2/. [Accessed 10.05.19].

[3] Hazard analysis and critical control points (HACCP) for the cannabis industry: Part 3. https://cannabisindustryjournal.com/feature_article/hazard-analysis-and-critical-control-points-haccp-for-the-cannabis-industry-part-3/. [Accessed 10.05.19].

[4] Hazard analysis and critical control points (HACCP) for the cannabis industry: Part 4. https://cannabisindustryjournal.com/feature_article/hazard-analysis-and-critical-control-points-haccp-for-the-cannabis-industry-part-4/. [Accessed 10.05.19].

[5] Part 117 current good manufacturing practice, hazard analysis, and risk-based preventive controls for human food. https://www.accessdata.fda.gov/scripts/cdrh/cfdocs/cfcfr/CFRSearch.cfm?CFRPart=117. [Accessed 09.28.19].

[6] Food Safety and Preventive Controls Alliance. Preventive controls for human food. 1st ed. 2016. p. 8-3.

[7] Food and Drug Administration. Warning letter. Sunshine International Foods, Inc.; 2019. https://www.fda.gov/inspections-compliance-enforcement-and-criminal-investigations/warning-letters/sunshine-international-foods-inc-580221-08082019 [Accessed 09.13.19].

[8] Food and Drug Administration. Warning Letter. Gourmet Catering to Go, LLC; 2020. https://www.fda.gov/inspections-compliance-enforcement-and-criminal-investigations/warning-letters/gourmet-catering-go-llc-596099-02032020 [Accessed 04.01.20].

[9] Food and Drug Administration, Center for Food Safety and Applied Nutrition, Office of Food Safety. Hazard analysis and risk-based preventive controls for human food: draft guidance for industry. Appendix 1: potential hazards for foods and processes. 2018.

[10] ImEpik LLC. PCQI online – FSMA training for preventive control qualified individuals, version 2. Risk matrix. 2019 [With permission].

[11] Colorado Department of Revenue, Marijuana Enforcement Division. Colorado marijuana laws and regulations. Charlottesville, Virginia: LexisNexis Matthew Bender & Company, Inc.; 2019.

[12] Bass G. FDA's perspective and experience on FSMA enforcement and inspections. International Association of Food Protection annual meeting, Louisville, Kentucky; 2019.

CHAPTER

Preventive controls

7

Ah, this is my happy place. This is what I do. The training from the Food Safety Preventive Controls Alliance (FSPCA) for the FDA Preventive Controls for Human Food (PCHF) rule and course curriculum for Preventive Controls Qualified Individual (PCQI) started at the end of 2015 [1]. I was in a cadre of lead instructors trained the end of January 2016, and I started delivering the course in the summer of 2016. In the first three years I delivered 29 face-to-face public or onsite workshops and trained 398 PCQIs. Over those years I fielded questions from those PCQIs when needed and worked with clients to identify and implement preventive controls. I wrote food safety blog posts on ConnectFood.com sharing my knowledge of and experience in compliance of the PCHF rule and updated a PCQI online curriculum at ImEpik. I guided clients through the writing of their hazard analysis and food safety plan or did the writing myself in concert with members of a company. Those three years are the same years that food businesses with more than 500 employees came under FDA enforcement, i.e., 2016, to small businesses in 2017 and very small businesses and fluid dairy plants coming under FDA enforcement in 2018. All food businesses are now under FDA enforcement of the PCHF rule. It has been an interesting time of industry preparing for compliance, FDA listening to industry questions on the rule, and the publishing of more FDA guidance on the Food Safety Modernization Act (FSMA) than anyone has time to read [1].

Good Manufacturing Practices (GMPs) are well established in the food industry and are the foundation on which preventive controls for food safety are built. In the writing of the hazard analysis by a food safety team, every process step and every ingredient are analyzed for biological, chemical, and physical hazards. After evaluating the risk of the hazard, the food safety team decides if the hazard is controlled by a GMP or preventive control. Controlling a hazard with GMPs is like a woman being released from prison and on probation. The probation program monitors her behavior and provides the rules for success. If the woman breaks the rules or commits another crime, she may be moved off probation and returned to prison. Similarly, if a hazard is controlled with GMPs, the GMP program monitors the control of the hazard. If the standard operating procedures (SOPs) and specifications are followed, the hazard remains controlled by a GMP. If the food safety team receives

Food Safety Lessons for Cannabis-Infused Edibles. https://doi.org/10.1016/B978-0-12-819512-3.00007-9

notice of a failure under the GMPs, the situation is reviewed and discussed, and the team may choose to move the control of the hazard to the more strict control of a preventive control.

Definition of a preventive control by the FSPCA [2]

> *... means those risk-based, reasonably appropriate procedures, practices, and processes that a person knowledgeable about the safe manufacturing, processing, packing, or holding of food would employ to significantly minimize or prevent the hazards identified under the hazard analysis that are consistent with the current scientific understanding of safe food manufacturing, processing, packing, or holding at the time of the analysis.*

When I teach the face-to-face PCQI course, I have my participants look at the title of the manual, *Preventive Controls for Human Food* [2]. I talk about the scenario of the participants returning to work with their PCQI certificate and having this conversation.

Coworker: "How was the training?"
PCQI: "Good."
Coworker: "What did you learn?"
PCQI: "Preventive controls and writing a food safety plan."
Coworker: "What are preventive controls?"
... crickets.

Although the FSPCA definition quoted earlier for preventive control is complex, it is very easy to describe what preventive controls are to someone: *preventive controls are procedures.* They are procedures that our employees and equipment do to achieve food safety. Employees follow written SOPs specific to a facility. When working as a consultant or instructor, I always start with the premise that the company is actually selling a safe product, unless I am working on a recall investigation in which a pathogen was found in the product or facility environment. Each food safety team will decide what procedures are preventive controls and will develop documents around that. Most food facilities have been operating safely for years or decades and now are getting caught up on documentation to prove the product is safe. Some food companies look at FSMA as another management program that will go away or be changed in some way. Those companies will be incorrect.

In contrast to the food industry, the cannabis industry has been undercover for decades or longer, and most cannabis companies and most cannabis-infused edibles manufacturers measure their time in the cannabis industry in units of months and years. Many cannabis-infused edible start-ups plan GMPs and preventive controls into a construction project and as part of edible safety management. In the same way a start-up food company would now be under FDA enforcement of FSMA rules from the get-go, cannabis-infused edible manufacturers are under state enforcement from the get-go with no lead time to leisurely implement edible safety. This is a different mindset from much of the food industry. The sense of urgency is strong in the cannabis industry compared to the food industry.

A cannabis-infused edible manufacturer may call the food safety team any team name that is appropriate; *edible safety team* was defined as a term in Chapter 4, Good Manufacturing Practice Compliance is Not Optional. A food safety plan is analogous to an *edible safety plan*, and food safety is analogous to *edible safety*. The roles for *supplier, receiving facility*, and *customer* were introduced in Chapter 2, Overview of the Cannabis Industry. Growing operations, drying operations, and extraction facilities are suppliers to edible manufacturers. The customer is the dispensary. The cannabis-infused edible manufacturing site is the receiving facility where preventive controls are applied to control hazards. This chapter will discuss the four types of preventive controls:

- allergen
- supply chain
- process
- sanitation

Important points about preventive controls are listed.

- *Preventive controls are procedures.*
- *A facility is not required to have a preventive control.* However, I dare you to show me a supplier who processes and sells a nonallergenic ingredient without biological or chemical hazards and without any physical hazards such as metal or glass. The closest I have come to finding such ingredient is liquid smoke packed in plastic drums, which are shipped out to another facility for filling of jugs and bottles.
- *A facility is not required to have at least one of each type of preventive controls.* There are four types of preventive controls, five types if "Other" is included.
- *A hazard may be controlled by one or more preventive controls.*
- *Every allergen, process, and sanitation preventive control must have written SOPs and records for monitoring, verification, correction, or corrective action. Verification follows monitoring.*
- *Each preventive control can be traced back to control a specific hazard identified in the hazard analysis.*

Allergen preventive controls

Important points about allergen preventive controls are listed.

- *Allergens are chemical hazards.*
- *Labeling of products with allergens in ingredients is an allergen preventive control.*

- *Effective cleaning of edible contact surfaces to remove allergen residues, called an allergen clean, is an allergen or sanitation preventive control.*
- *The FDA PCHF rule does not require the validation of an allergen clean, but a company should validate the allergen clean.*
- *A receiving facility should audit suppliers to determine the risk of cross-contact of allergens present at a supplier's facility and not present in the receiving facility's products.*
- *A written allergen management program will include SOPs for allergen control using both GMPs and preventive controls.*

Step-by-step directions for the identification of allergen preventive controls

The regulation for allergen preventive control is found in the FDA PCHF rule Subpart C [4]. Allergens must be controlled at your suppliers, at the receiving facility, and with compliant labeling of your product. When my clients are performing a hazard analysis on their list of ingredients and process steps used to make an edible or group of edibles, i.e., the product, in a single edible safety plan, I recommend the following steps.

1. Start with the master list of ingredients for all products in a single edible safety plan. Create an allergen matrix for individual ingredients noting which of the FDA big eight allergens are naturally present in the ingredient [5], including ingredients with no allergens. If an ingredient is purchased from more than one supplier, then list the ingredient multiple times until all suppliers are listed. Cannabis-infused edible manufacturers outside the United States must seek out their government's rule for allergen declaration on the label, as different countries have different requirements.
2. Every product with an ingredient in which an allergen is naturally present must be correctly labeled as required by the Food Allergen Labeling and Consumer Protection Act [5].
   a. I recommend an allergen preventive control at Receiving of preprinted labels. Write and implement a Receiving of Labels SOP for holding and inspecting label information at Receiving.
   b. I recommend an allergen preventive control of monitoring of the label at Packaging for the product. A common error is to pack the product into the wrong primary packaging. If the product goes into commerce and the label is wrong but there is no hazard, the product is misbranded. If the allergen labeling is wrong, the product is adulterated, and a Class I recall may result.
3. List all the suppliers, i.e., vendors, of all the ingredients in the master list with and without allergens in the ingredients. Implement a procedure as part of a supplier approval SOP for the disclosure of allergens present in the facility of the supplier. This step does not apply to a distribution center, and the receiving facility must go back enough steps to get to the supplier who packages the

ingredient. Document allergens that are present at the facility of the supplier in addition to the allergens naturally present in the ingredient. All suppliers must verify that the hazard of cross-contact of undeclared allergens was controlled. When other allergens are present at the supplier and there is the potential for cross-contact,

   a. the receiving facility must audit the supplier to determine the risk of cross-contact, and
   b. the management of the receiving facility may choose precautionary labeling. Precautionary labeling is unregulated and often misinterpreted.

4. Sort the master list of ingredients into (1) ingredients received with naturally present allergens and (2) ingredients received without naturally present allergens. The edible safety team is finished with the latter list, if there is control of cross-contact at the supplier per step 3. The list of ingredients received with naturally present allergens carries on for control.
5. Write and implement an allergen management program. Control of allergens starts at Receiving of ingredients with naturally present allergens and follows the ingredient at every step in the facility with labeling, color-coding, dedicated utensils, cleanup procedures, and more. The allergen management program includes employee training on allergens. Many of the procedures in the allergen management program can be handled as GMPs and not as preventive controls.
   a. The edible safety team will decide if a procedure in the allergen management program is a GMP or an allergen preventive control.
6. Next write every step from Receiving to Shipping on the hazard analysis form. Identify process steps where there are edible contact surfaces, and note where there is the potential of allergen cross-contact resulting in the hazard of undeclared allergens. Once the package is sealed around the product, the potential for cross-contact theoretically stops. Using the risk matrix as a tool and in team discussion decide if the hazard is controlled in the allergen management program by a GMP. If not, control of the hazard requires an allergen preventive control. If the control of cross-contact is with an allergen clean of edible contact surfaces at a process step, the preventive control may be labeled as either a sanitation preventive control or an allergen preventive control. In this case, the type of preventive control does not matter, only that the hazard of undeclared allergen is being controlled. An allergen clean uses cleaning procedures to remove allergen residue from edible contact surfaces (Box 7.1).

Box 7.1 Important points about cleaning and sanitizing

Cleaning: The purpose of cleaning is to remove food residues, including mineral deposits, from food contact surfaces. Cleaning of non-food contact surfaces in the manufacturing environment prevents cross-contact and cross-contamination to the food contact surfaces.
Sanitizing: The purpose of sanitizing is to kill spoilage-causing or pathogenic microorganisms. Sanitizing always follows cleaning as a separate step and is not effective in combination with cleaning.
Sanitation: A generic term inclusive of cleaning and sanitizing.

Application to cannabis-infused edibles

Potential hazards requiring an allergen preventive control include the following.

Undeclared allergen in ingredient from supplier: The edible manufacturer's supplier approval program may include an inventory of allergens present at the supplier's facility and may include an audit of the supplier's facility to observe the prevention of cross-contact. For most edible manufacturers, this step is insurmountable because ingredients are purchased from retail or wholesale distributors. Document due diligence in seeking this information from the supplier and through the distributor. A good example is purchasing of a single type of nut; nuts generally are roasted and packaged in facilities that process other nuts, and there is the potential for cross-contact of one nut allergen into a different nut. The edible manufacturer should go back to the nut processor to obtain information on additional allergens in the facility.

Undeclared allergen on label of finished product: The edible manufacturer creates a label that declares the known allergens.

Cross-contact of allergen into finished product without the allergen: The edible manufacturer schedules production based on the introduction of allergens and schedules an allergen clean after the production of an edible with a unique allergen.

Supply chain preventive controls

Important points about supply chain preventive controls are listed.

- *Supply chain preventive controls are for ingredients, not process steps.*
- *A receiving facility identifies a supply chain preventive control when a hazard is controlled at the supplier and not at the receiving facility.*
- *The supply chain preventive control for an ingredient is always controlled at Receiving, and "Yes" is checked in the last column of the hazard analysis form, indicating the location of the control* (Fig. 7.1).
- *Most chemical hazards are controlled by the supplier,* e.g., *undeclared allergens, radioactive chemicals, mycotoxins, pesticides, drug residues, industrial chemicals, decomposition by-products, and unapproved food or color additives.*
- *Include water as an ingredient in the hazard analysis for the risk assessment of the receiving facility's water.*

Supply chain preventive controls for ingredients require evidence of safety through the collection of documentation. The regulation for supply chain preventive control is found at the end of the FDA PCHF rule in Subpart G [6]. Why is this preventive control separate from the others in Subpart C? Simple. There is no physical work onsite at a receiving facility for a supply chain preventive control. A supply chain preventive control is all about verification through the collection and reviewing of documentation by the receiving facility and from the supplier; there is no

Hazard Analysis								
(1) Ingredient / Processing Step	(2) Identify potential food safety hazards introduced, controlled or enhanced at this step		(3) Do any potential food safety hazards require a preventive control?		(4) Justify your decision for column 3	(5) What preventive control measure(s) can be applied to significantly minimize or prevent the food safety hazard? *Process including CCPs, Allergen, Sanitation, Supply-chain, other preventive control*	(6) Is the preventive control applied at this step?	
			Yes	No			Yes	No
	B							
	C							
	P							

FIGURE 7.1

Complete column 5 in the hazard analysis form with one of the four types of preventive controls [3] or "Other." Separate records will follow the hazard analysis and detail the procedures for monitoring, verification, correction or corrective action, and validation.

monitoring at a receiving facility. In the same way that a receiving facility receives an ingredient, the receiving facility receives documentation. What about audits? Yes, but the onsite audit is at the *supplier*, not at the receiving facility, and an audit report is generated as documentation. There is minimal creation of records at a receiving facility, and most of these records will be associated with Receiving of an ingredient with a supply chain preventive control. Examples of supply chain preventive controls are in the Application to cannabis-infused edibles section following the step-by-step directions.

Step-by-step directions for the identification of supply chain preventive controls

When my clients are performing a hazard analysis on their list of ingredients used to make the product in a single edible safety plan, I recommend the following steps.

7. Start with the master list of ingredients for all products in a single edible safety plan and a corresponding single hazard analysis.
8. Sort the master list into (1) ingredients with no hazards at Receiving, e.g., salt or vinegar, and (2) ingredients with hazards requiring a preventive control without the type of preventive control noted. On the hazard analysis indicate those ingredients with no hazards, and the edible safety team is finished with that list. Only the second list carries on for assignment of preventive control. The second list includes all the ingredients with hazards where the hazard is controlled at the supplier, at the receiving facility, or by the customer.

9. Separate the list of ingredients from step 8 into receiving as ambient, refrigerated, and frozen.
   a. As an option with any of the three receiving temperatures, a facility may group ingredients together, for example, by the type of packaging, e.g., canned goods; by allergen, e.g., milk; or by type, e.g., dried spices.
   b. Ingredients may be analyzed in groups only if all the ingredients in the group have the same hazards. If unique hazards are identified, it may be best to remove the ingredient from the group and address the ingredient individually.
10. The edible safety team names the biological, chemical, and physical hazards for every group of ingredients or individual ingredient in column 2 of the hazard analysis form. Some hazards will be controlled by GMP programs. I encourage clients to manage low-risk hazards with GMP programs such as pesticide management, pest management, and water quality and not assign a preventive control. *If it ain't broke, don't fix it* [7]. It is only when there is evidence of a hazard through failed tests, recalls, or corrective action at the supplier, and the hazard was not controlled by a GMP program, that a preventive control is required.
11. The edible safety team agrees on which hazards require a preventive control.
   a. Column 3 of the sample hazard analysis form is checked as Yes for requiring a preventive control; see Fig. 7.1.
   b. If during analysis the edible safety team finds an ingredient without a hazard then that ingredient moves to the list of ingredients with no hazards.
12. For every hazard requiring a preventive control, the edible safety team determines the entity responsible for the control of the hazard as
   a. the supplier;
   b. the receiving facility, i.e., cannabis-infused edible manufacturer; or
   c. the customer or consumer.

Control of hazards by the customer or consumer may be identified with the "Other" preventive control. Examples include hot-chocolate mix and tea made with hot water or dough baked either in a foodservice facility or by the consumer in their home oven. Cannabis companies selling an edible with the biological hazard controlled by the customer or consumer must provide validated cooking instructions. Having the customer control the hazard is very tricky, and I recommend that cannabis companies stay away from edibles which require the customer to control the hazard. Continue the steps from step 12 above.

13. Using the list of ingredients where the supplier controls the hazard and for each ingredient or group of ingredients, label the type of preventive control as supply chain preventive control at Receiving of the ingredient. This completes the hazard analysis form for the ingredients or group of ingredients. Remaining ingredients where the hazard is controlled by the receiving facility will be addressed in the Process preventive controls section.

14. If an ingredient requiring a supply chain preventive control is purchased from more than one supplier, a supply chain preventive control is applied to all suppliers. All suppliers must verify that the hazard was controlled.

Currently, food contact packaging material is not required to be evaluated for a supply chain preventive control. Manufacturers should follow due diligence in the purchasing of packaging material without hazards that could migrate to the edible and should take measures to store and handle packaging material to prevent cross-contact and cross-contamination.

Application to cannabis-infused edibles

Potential hazards requiring a supply chain preventive control are discussed in the following. Remember supply chain preventive controls are applied to ingredients.

Biological

Clostridium botulinum: *Clostridium botulinum* produces a deadly toxin in hermetically sealed packaging such as cans or pouches and where the acidity is above pH 4.6. The edible manufacturer should purchase commercially available products and should not use high-risk, home-canned ingredients. *C. botulinum* spores may survive inadequate thermal processing and grow in sealed packaging. Some concern of botulism from concentrates exists in the cannabis industry. The characteristics of concentrates makes the growth of *C. botulinum* and the cause of botulism very unlikely in the cannabis ingredient for edibles. The extraction process removes water, resulting in a concentrate with not enough available water for growth. Acidity of the concentrate below pH 4.6 or refrigeration of the concentrate prevents the growth of *C. botulinum*. A combination of low water activity, high acidity, and refrigeration to prevent clostridial growth may be verified in a challenge study of a concentrate.

Vegetative pathogens: *Salmonella*, pathogenic *Escherichia coli*, and *Listeria monocytogenes* are vegetative pathogens easily killed by cooking steps such as roasting and baking. Manufacturers without a kill step for an edible must verify the control of biological hazards for each ready-to-eat (RTE) ingredient in the edible. An example of an RTE edible without a kill step would be a protein bar for which the ingredients are blended, the bar is formed, and there is no baking step. Roasted nuts and chocolate are examples of ingredients in which the control of the biological hazard may be with a supply chain preventive control.

Chemical

Undeclared allergens: See the Allergen preventive controls section.

Pesticides, heavy metals, and residual solvent: Follow the state code for testing of the cannabis concentrate as an ingredient. The team should evaluate the need for additional pesticide testing based on risk and supplier compliance with state code, as the state code varies from state to state. For noncannabis ingredients manage pesticides or heavy metals with a GMP program. Examples of ingredients that may introduce additional pesticides or heavy metals include fruits, vegetables, grains and grain products such as flours. A pesticide management program for noncannabis ingredients may control the hazard of pesticides with a GMP program and without a preventive control.

Radioactive chemicals: Obtain your local water testing report to verify the safety of water used as an ingredient and used for sanitation of equipment and processing rooms. Water safety is controlled as a GMP program unless water tests indicate unsafe water.

Drug residues: Determine the risk in honey and all animal products such as dairy products or lard.

Mycotoxins: Follow the state code for testing of the cannabis concentrate as an ingredient. The edible safety team should evaluate the need for additional mycotoxin testing of concentrate beyond the state code and based on risk and supplier compliance, as the state code varies from state to state. Determine the risk of mycotoxins in the noncannabis ingredients of flour, additional grain products, and nuts. The production of mycotoxins is controlled by low water activity that prevents the growth of mold.

Industrial chemicals, decomposition by-products, and unapproved food or color additives: Determine the risk of other chemical hazards at the supplier.

Physical

Suppliers are expected to send ingredients with no detectable physical hazards. When physical hazards are observed follow your corrective action procedure and document the results.

Process preventive controls

Important points about process preventive controls are listed.

- *Process preventive controls are critical control points in a Hazard Analysis and Critical Control Points (HACCP) plan.*
- *All process preventive controls must be validated.*
- *Metal detection is a process preventive control.*
- *If a process step has been validated, it may be a process preventive control.*
- *Annual validation is not required for regulatory purposes but may be required for certification.*

Step-by-step directions for the identification of process preventive controls

Most process preventive controls are designed to kill pathogens. The receiving facility will control the biological hazards coming in with the ingredients, not the supplier, using an appropriate combination of time and temperature for the kill step. Examples of kill steps include cooking and baking. Nonthermal parameters for kill steps are also found in the food industry and include acidification, concentration of preservatives, and high-pressure processing. The regulation for process preventive control is found in the FDA PCHF rule Subpart C [4]. When my clients are performing a hazard analysis on their list of ingredients and process steps used to make the product in a single edible safety plan, I recommend the following steps.

15. For every ingredient requiring a preventive control at the receiving facility in step 12b mentioned earlier, the edible safety team reviews the Receiving step of the ingredient. It is important to name the hazard at Receiving of the ingredient.
16. If the receiving facility will control the **biological hazard** in a process preventive control after Receiving, "No" is checked in column 6 because the control is not at this step, i.e., the control is not at Receiving but at a later step (Fig. 7.1).
17. The edible safety team finds the kill step in the hazard analysis among the process steps, labels the kill step as a process preventive control, and checks "Yes" in column 6 for the control is at this step, i.e., the kill step.
18. The edible safety team repeats the review for **chemical hazards**. While most chemical hazards are controlled by the supplier and at Receiving, the receiving facility may control the hazards of radioactive chemicals and industrial chemicals in their water, onsite decomposition by-products, and onsite unapproved food or color additives. These hazards are often controlled by GMPs and not a preventive control.
19. The edible safety team repeats the review for **physical hazards** starting at Receiving and evaluating the risk at each process step. It is common to control for the hazard of detectable metal with the use of a metal detector as a process preventive control. A process step in which equipment has metal-on-metal contact can generate the hazard of metal. Without a metal detector, the facility may control for the hazard of metal by inspecting the processing equipment to monitor that parts are intact after processing and no metal has been included in the edible.
20. If the facility fills glass containers and notes the physical hazard of glass, a glass breakage procedure can be handled as a GMP.

Application to cannabis-infused edibles

Potential hazards requiring a process preventive control are discussed.

Biological

Vegetative pathogens: Validate a kill step, e.g., roasting, baking, or cooking. Monitor time and temperature as appropriate.

Chemical

Undeclared allergens: Prevent cross-contact and create label checks for allergens; see the Allergen preventive controls section.

Unapproved food or color additives: Manage ingredient addition as a GMP with the recipe batch sheet.

Incidental cleaning agents, sanitizers, or lubricants: Manage the security and use of chemicals within the facility as a GMP.

Physical

Metal: A process preventive control is required for the hazard of metal at each process step with metal-on-metal contact. Short of a metal detector, small businesses can do visual inspection after production to monitor if equipment surfaces are intact. Visual inspection is limiting in comparison to a metal detector because a nut or bolt that has fallen from equipment or from the processing room into the edible may go undetected.

A warning to cannabis-infused edible manufacturers

Stay away from cannabis-infused, cold-brewed beverages! This is a microbial nightmare because there is no kill step. Beware of current fads in food consumption and nontraditional products and processes.

Sanitation preventive controls

I am a huge proponent of cleaning and sanitizing procedures being extremely important to the success of edible safety. My dissertation at the University of Minnesota addressed cleaning and sanitizing of ultrafiltration membranes, I have been on my soapbox ever since about sanitation, and I have a lot to say about cleaning and sanitizing. I believe sanitation crews, i.e., those who clean and sanitize the utensils, equipment, and processing environment, should work on first shift in a facility that runs 20–24 hours and be highly paid. I shifted my paradigm to believe that production follows only when the sanitation crew has successfully and thoroughly finished their work, not that sanitation starts when production finishes. Most food facilities are 24/7 with employees working around the clock. Cannabis-infused edible manufacturing facilities that do not run for 24 hours must resanitize equipment that is enclosed and has sat idle for more than four hours before use of the equipment. Utensils and open equipment that quickly dry out may not require resanitizing before use, but it wouldn't hurt.

The cannabis-infused edibles industry will learn from the food industry about cleaning and sanitizing procedures and monitoring of food contact surfaces (FCSs) for both allergens and biological hazards. The knowledge held by the food industry is immense and will be shared readily. Cannabis-infused edibles are primarily shelf-stable RTE at this point in history, although I learned of a cannabis-infused ice cream manufacturer in Massachusetts. Ice cream is a frozen RTE product. A sanitation preventive control is required for the cleaning and sanitizing of FCSs and their environment when the product is RTE [4]. GMPs may be appropriate for the cleaning and sanitizing of non—food contact surfaces (NFCSs) not located in proximity to the product.

Important points about sanitation preventive controls are listed.

- *For an allergen clean, the FDA PCHF rule does not require validation, but a company should validate the allergen clean.*
- *For effective sanitation, the FDA PCHF rule does not require validation, but a company should validate sanitation.*
- *Facilities producing RTE products must have an environmental monitoring program (EMP) for pathogens of concern.*

Step-by-step directions for the identification of sanitation preventive controls

The regulation for sanitation preventive control is found in the FDA PCHF rule Subpart C [4]. When my clients are performing a hazard analysis on their process for a single edible safety plan, I recommend the following steps.

21. Using a process flow diagram and walking the line(s) with the edible safety team confirm whether all points of potential allergen cross-contact have been identified and addressed as in step 6.
22. Using a process flow diagram and walking the line(s) identify the post-kill steps, i.e., the location where the product leaves the kill step, where the product is potentially exposed to an environmental pathogen through to the primary packaging step. Examples include when a bakery product leaves the oven or after the gummy syrup is portioned into the mold. Identify the hazard of environmental pathogen from the post-kill location until the primary package is sealed.
23. Write and implement sanitation SOPs and create corresponding records for documentation.
24. Review and revise, if needed, the *master sanitation schedule* for the sanitation of all parts of the facility to control cross-contact and cross-contamination.
25. Review the *important points about preventive controls* mentioned earlier. *Each preventive control can be traced back to controlling a specific hazard identified in the hazard analysis.* If the cleaning or sanitizing procedure is not directly controlling a hazard as identified in the hazard analysis, the procedure is not in support of a preventive control. For example, cleaning of the employee break rooms and restrooms is a GMP, not a preventive control. Assess if sanitation SOPs will be part of GMP sanitation programs or preventive controls. Many of the sanitation SOPs can be managed as GMPs and not as preventive controls.
26. Determine the pathogens of concern by using resources. At a minimum the following should be performed:
    a. If the processing and packaging environment is wet and cleaned with traditional wet sanitation, the pathogen of concern is *L. monocytogenes*.

Most cannabis-infused edibles are not from a wet operation. An example of a wet operation is cannabis-infused beverages. However, the small size of cannabis-manufacturing kitchens and the application of traditional wet cleaning and sanitizing make the edible and processing environment at risk for *L. monocytogenes*. I recommend that *L. monocytogenes* be identified as a hazard in the wet areas of an edible facility.

b. If the processing and packaging environment is dry, the pathogen of concern is *Salmonella.* Most cannabis-infused edibles are from dry operations. Examples of dry operations are chocolates, soft confections, hard confections, baked goods, and pressed pills and capsules. I recommend that *Salmonella* be identified as a hazard in the post-kill step dry areas of an edible facility.

c. Identify additional pathogens of concern in the state code.

27. Write and implement an EMP for the pathogens of concern in RTE products. An EMP is required of RTE food manufacturers [8], and expect FDA to hold RTE edible manufacturers to the same requirement as RTE food manufacturers. Chapter 8 provides more details on an EMP.

Application to cannabis-infused edibles

Potential hazards requiring a sanitation preventive control are discussed.

Biological

Environmental pathogens: Sanitizing of edible contact surfaces follows cleaning, and its purpose is to kill pathogens that remain after cleaning. The equipment, utensils, and rooms for the post-kill steps must be cleaned and sanitized to prevent cross-contamination of an environmental pathogen into the product. In the future, cannabis-infused edible manufacturers can expect FDA to require implementation of a robust EMP as is the requirement for the food industry.

Chemical

Undeclared allergens: The purpose of cleaning edible contact surfaces is to remove product residues, including allergens. See the Allergen preventive controls section.

Cleaning agents and sanitizers: Implement sanitation SOPs for the secure storage and use of cleaning agents and sanitizers throughout the facility as a GMP program.

Important points about an EMP are listed.

- *The EMP is a verification activity of a sanitation preventive control.*
- *The hazard analysis identifies the hazard of environmental pathogens from the location of a post-kill step to sealing of the primary packaging for an RTE edible.*
- *With no kill step, the EMP starts at batching where each ingredient is first exposed to the environment at the receiving facility.*

Sanitation preventive controls go hand in hand with EMPs. Chapter 8, Nobody is Talking About Environmental Monitoring, continues the discussion of verification of a sanitation preventive control with chemical indicators, microbial indicators, and pathogen testing. Sanitation of processing and packaging rooms, equipment, and FCSs is the primary control for environmental pathogens, and the importance of sanitation cannot be overemphasized. Built into the manufacturing of a safe RTE edible is sanitation preventive controls to control the hazard of environmental pathogens after the kill step. In most food facilities, this means the equipment is shut down from production and packaging and both solid waste and packaged product are removed from the processing and packaging areas. If it is a wet operation, cleaning and sanitizing are typically done daily. In the food industry, dry operations are cleaned less frequently and with unique waterless techniques. Most RTE edible manufacturing and packaging areas are classified as a dry operation. For example, an oven for baking would not be cleaned daily; the frequency of oven cleaning would be what is reasonable and indicated on the master sanitation schedule.

However, because edible processing facilities are called kitchens and many states regulate cannabis kitchens as restaurants, there is an understanding that equipment such as bowls, scoops, and trays will be washed, sanitized, and dried daily, which is appropriate for small-scale operations. Most cannabis kitchens will have a three-compartment sink to clean and sanitize. The washing and sanitizing of parts in the three-compartment sink is a concern if the splashing of water creates a wet processing environment for the proliferation of both *Salmonella* and *Listeria* in an area that is part of or proximate to a dry operation such as baking and packaging. It may be disastrous with regard to environmental pathogens to have a three-compartment sink in the same room as processing and packaging. *Listeria* is associated with wet environments such as drains. Creating additional wet environments is conducive to *Listeria* proliferation outside the drain. *Salmonella* can lie dormant in a dry operation until water comes along when it springs to life, growing and multiplying.

Effective cleaning and sanitizing will physically remove the residues of the edible and chemically kill any remaining pathogens hiding in microbial niches. It is not possible to have effective cleaning and sanitizing as a single step, and the sanitation standard operating procedure (SSOP) should follow the sequence of rinse, clean, rinse, and sanitize. If it is a dry operation, the parts must be completely dry before use. In my experience with the manufacture of spray-dried milk as an ingredient, the system is enclosed, and companies would be ecstatic if they could never shut down to clean. Spray dryers are only stopped for maintenance and repair; after maintenance the spray dryer is cleaned and sanitized. Before starting production and sending the fluid to be dried to the spray dryer, it is not unusual for the facility to run the spray dryer with just hot air continuously for more than 24 hours to dry out the interior! The presence of residual water from cleaning and sanitizing is dangerous for the potential of *Salmonella* to multiply on the inside of equipment, pipes, and pumps.

Hygienic design and coupling of equipment are critical for cleanability. FCSs must be maintained in good condition for complete cleaning and sanitizing. Over time, gaskets will wear and metal will pit, corrode, or become scratched. Imagine a wooden cutting board with scratches. Pathogens get to the bottom of the scratch. The bottom of the scratch to the pathogens is in proportion to you and me standing at the bottom of the Grand Canyon and looking up. With every use of the cutting board, food builds up on top of the pathogens like an avalanche. After adequate cleaning, the role of a sanitizer is to penetrate any remaining layer of food residue and kill the pathogens that were not removed during cleaning. Without cleaning first, sanitizers will not be effective in killing pathogens.

How are FCSs adequately cleaned? There are two types of cleaning. The three-compartment sink is an example of clean-out-of-place (COP), meaning the parts are taken to the sink for washing and sanitizing. Another type of COP is a long tank that is filled high enough to cover the parts. By circulating the tank and injecting steam into the detergent flow, parts can be washed at higher temperatures with greater agitation for cleaning than can be achieved by manual cleaning. Separately, this type of tank can be used for soaking parts in sanitizer for an extended time. The second type of cleaning is clean-in-place (CIP). CIP is for connected equipment, pipes, and pumps, such as the enclosed system of the spray dryer. The system is cleared of product and thoroughly rinsed. Detergent is circulated through the system with agitation, at high temperature, and for an extended time. The detergent is metered into the system and the concentration is monitored. After the system is drained and rinsed, the system is CIP sanitized.

Following GMPs for the integrity of the facility is critical. The roof must be repaired to keep water out. The materials of the ceiling, walls, floors, and drains must be kept in good condition and cleanable. The same principles apply for cleaning of the processing and packaging rooms as for equipment, i.e., remove debris from the room, rinse, clean, rinse and sanitize. When a sanitation preventive control is required for the prevention of cross-contact of allergens or for eliminating pathogens in the processing environment, a facility will monitor and verify cleaning and sanitizing procedures.

Monitoring cleaning and sanitizing

In monitoring of cleaning and sanitizing, separately, a monitoring record is created to record the date, time of day, chemical and its amount, temperature, operator, and length of cycle or any other parameter that is important for effective cleaning and sanitizing. The SSOP will detail the procedure; the record will prove the procedure was followed. Any deviation from the SSOP initiates a ***correction*** by the operator to restore the system back into control.

Verification of cleaning and sanitizing

Record review

Record review is a verification activity. A qualified sanitation supervisor or manager reviews, signs, and dates the monitoring record. In this process, monitoring is being verified, meaning (1) complete information was recorded and (2) the data show the hazard is being controlled. Any deviation from the SSOP initiates ***corrective action***.

Use of dip strips

Dip strips are used to test the concentration of sanitizer. When preparing sanitizer, if the operator records the amount of concentrated sanitizer and amount of water for diluting, then that is a monitoring record. The operator is making a record that shows the SSOP was followed. Theoretically, a record review would verify the sanitizer was prepared correctly and dip strips would not be needed. The dip strip measures the concentration of the sanitizer and that the sanitizer was prepared correctly or has maintained its strength. Any deviation initiates correction if the sanitizer has not yet been used, and a fresh solution is prepared. Additional verification is record review by a supervisor or manager. If the dip strip is used before the use of the sanitizer, the step could be labeled monitoring. If the dip strip is used to check if the sanitizer has maintained its strength through use, the step could be labeled verification.

Real-time swabbing

Following an allergen clean, many companies swab an FCS to verify cleaning was effective. There are many options on the market for protein or allergen swabs that give almost immediate results at the site, i.e., real-time results. If the test fails, the correction is that the sanitation crew recleans. Adenosine triphosphate (ATP) swabs verify cleaning in real time by indicating the presence of cells, any type of cell. If the test fails, the correction for the sanitation crew is to reclean. Like the protein and allergen swabs, ATP swabs give almost immediate results at the equipment site. The protein, allergen, and ATP swabs are methods for chemical analysis. In contrast to real-time swabbing, swabs of sample sites can be tested for microbial indicators or pathogens. The growth of microorganisms takes time, and laboratory results will be reported days later after swabbing.

Implementation of a sampling plan

A sampling plan is a compilation of documents detailing the collection and testing of surface samples for monitoring and verification. Once the sampling plan for environmental monitoring is developed, SOPs detail the procedures for supplies, sampling procedure, and preparation of samples for transportation to the laboratory. In my experience, an effective sampling plan is constantly evolving to meet the needs of the facility.

Training and refresher training

Sampling plan SOPs serve as training tools and a means to verify all technicians are following the same procedures. Supervisors should frequently observe the sampling procedure by technicians as a verification of the EMP. When there are deviations from the sampling plan and when positive results for the presence of pathogens are reported, supervisors should take the opportunity to meet with staff to discuss the results and listen for improvements to the procedures and overall EMP.

Choosing the supplies

Surface samples are collected with sterile swabs or sponges of a site or a predefined area. Technicians must be provided with supplies to collect samples aseptically, i.e., without contaminating the sample during the sampling procedure.

When are surfaces swabbed?

This question speaks both to the time of day and the frequency, which is confusing, and the food industry struggles with the answers. If the purpose of swabbing is to verify cleaning, then sites must be swabbed after cleaning and before sanitizing. This is best done with ATP swabs, giving real-time results. If the purpose of swabbing is to verify sanitation, then sites must be swabbed after cleaning and sanitizing and before production or the use of equipment. This is best done with the supply of a sponge-on-a-stick and microbial indicator tests for sites that are FCSs. Samples from FCSs should not be tested for pathogens because a positive result for pathogen presence would implicate the day's production from that equipment. If the purpose is to verify the absence of pathogens in the processing environment, then NFCS sites should be swabbed after three hours into and during production. This is best done with the supply of sponge-on-a-stick and appropriate pathogen tests. For the latter sites, some facilities may choose to swab at the end of production for the day. Before taking a sample and testing for a pathogen or pathogens think through the implication of a positive result for pathogen presence on the day's entire production. As most edibles are not released for sale until laboratory results are reported, the cannabis industry is positioned to follow a hold and test program, which is an advantage the cannabis industry has forced upon it by the state code (Box 7.2).

> **Box 7.2 Equipment sanitation before use**
>
> By the way, if equipment stands for more than four hours after cleaning and sanitizing, it is recommended that the equipment is resanitized before use. This best practice exists for enclosed parts of the equipment that may remain damp, as they are not exposed to the environment and do not dry completely after sanitizing. If utensils, tools, or parts are cleaned, sanitized, allowed to dry, and stored without cross-contamination, the recommendation to resanitize may not apply.

Transportation of samples to the cannabis laboratory

After the collection of site samples, a technician will follow an SOP to pack both the samples and paperwork for transportation to the cannabis laboratory. Concentrate and edibles may be included in the shipment with environmental samples. The technician must demonstrate attention to detail in making the shipping manifest for accuracy and security of samples. The state code rarely allows for the cannabis company employee to transport the samples, and instead, an employee of the cannabis laboratory commonly goes to the cannabis facility to transport the samples back to the laboratory. States require licensing of cannabis laboratories, and manufacturers are generally limited in the choice of laboratories because samples must stay within a state. Courier service to local laboratories across state lines, e.g., an Indiana facility to an Illinois laboratory, and overnight shipping for next day delivery is common in the food industry, but illegal in the cannabis industry. For example, I had an environmental risk project for evaluating the processing environment of a facility in which I had 100 environmental site samples shipped to a California laboratory for microbial testing from Kentucky by overnight delivery. If the samples were cannabis, the shipping would be illegal. Although it is common in the food industry to cross state lines with samples for testing, it is a felony in the cannabis industry. The cannabis industry can look forward to the post-descheduling world in which samples can be shipped with a commercial carrier to any laboratory of their choosing. The cannabis laboratory business and pricing of tests will get highly competitive.

References

[1] FSMA rules and guidance for industry, guidance for industry & others. https://www.fda.gov/food/food-safety-modernization-act-fsma/fsma-rules-guidance-industry#Guidance. [Accessed 09.13.19].

[2] Food Safety and Preventive Controls Alliance (FSPCA). Preventive controls for human food. 1st ed. 2016. p. 1–7.

[3] Food Safety and Preventive Controls Alliance (FSPCA). Preventive controls for human food. 1st ed. 2016. p. 8–3.

[4] FSMA final rule for preventive controls for human food. https://www.fda.gov/food/food-safety-modernization-act-fsma/fsma-final-rule-preventive-controls-human-food. [Accessed 09.13.19].

[5] Food and Drug Administration, Food allergen labeling and consumer protection act. https://www.fda.gov/food/food-allergensgluten-free-guidance-documents-regulatory-information/food-allergen-labeling-and-consumer-protection-act-2004-falcpa. [Accessed 09.13.19].

[6] Hazard analysis and risk-based preventive controls for human food: draft guidance for industry, Appendix 15: Supply-Chain Program for Human Food Products. https://www.fda.gov/media/110443/download. [Accessed 09.13.19].

[7] The phrase finder. https://www.phrases.org.uk/meanings/if-it-aint-broke-dont-fix-it.html. [Accessed 09.13.19].

[8] Control of *Listeria monocytogenes* in ready-to-eat foods: guidance for industry draft Guidance. https://www.fda.gov/media/102633/download. [Accessed 10.14.19].

CHAPTER

Nobody is talking about environmental monitoring

8

Environmental monitoring

It scares me that the cannabis industry does not have environmental monitoring in its vocabulary for edibles. Pathogens will enter the edible facility and should be controlled with a combination of Good Manufacturing Practices (GMPs) and preventive controls, including a validated kill step as appropriate and effective sanitation. Kill steps are designed to reduce pathogens by five or six log (99.999% or 99.9999%). The environmental monitoring program (EMP) is facility-specific for the containment of pathogens after the post-kill step area and uses both indicator microorganism and pathogen testing for verification. As the cannabis-infused edible industry learns from the food industry and applies best practices such as an EMP, a strong edible safety culture will develop in the prevention of cross-contamination. Environmental monitoring seeks to find environmental pathogens, and, when found, intensive corrective action will remove or destroy the environmental pathogen and prevent cross-contamination into the edible. Pathogen testing plays a role in finding environmental pathogens from samples collected from non–food contact surfaces (NFCSs).

Environmental monitoring is a verification activity for a sanitation preventive control within an edible facility when the edible is ready-to-eat (RTE) and applies to the cannabis industry in that most edibles are RTE, including the bakery items of cookies, bars, and brownies; chocolates; confectionary items of lollipops, lozenges, mints, and gummies; beverages; tablets or pills; and tinctures. The hazard analysis for an RTE edible identifies the biological hazard of environmental pathogens. The purpose of an EMP is to control the hazard of an environmental pathogen within an edible facility and prevent cross-contamination of the edible with a pathogen found within the facility environment. Primary control of the hazard lies with effective cleaning and sanitizing of equipment, utensils, and rooms. Environmental monitoring applies hygienic zoning and starts with the documentation that equipment, utensils, and rooms for processing and packaging are visually clean and incorporates the use of indicator tests on swab samples of food contact surfaces (FCSs) and pathogen tests on swab samples from NFCSs. Environmental monitoring is not appropriate for facilities manufacturing non-RTE edibles that will receive a kill step applied by the customer or consumer, such as cooking or baking.

Food Safety Lessons for Cannabis-Infused Edibles. https://doi.org/10.1016/B978-0-12-819512-3.00008-0

Where is environmental monitoring conducted?

When a microbiologist talks about monitoring the environment, the location of that environment is primarily within the walls of edible manufacturing and packaging rooms and not inclusive of growing or drying of cannabis. Edible manufacturers start with a cannabis concentrate as an ingredient and must be separate from the operations of growing, drying, and extraction to control pathogens in the edible environment. The primary area of concern is after a kill step has destroyed pathogens, through processing steps where the edible is exposed to the environment, and until the edible is sealed in its primary packaging.

The rooms and equipment within must be cleaned and sanitized as part of both GMPs and sanitation preventive controls listed in edible safety plans for RTE foods. RTE food facilities are required to implement an EMP as presented in both Chapters 6, The Hazard Analysis, and Chapter 7, Preventive Controls, in the sections on sanitation preventive controls. It is only a matter of time before RTE edibles facilities will be required to implement an EMP.

Kill steps of cooking, roasting, baking, or pasteurization easily destroy vegetative pathogens, i.e., nonsporeformers such as *Salmonella* and *Listeria monocytogenes*. An EMP seeks out environmental pathogens that may enter the facility, travel to the manufacturing and packaging rooms, and cross-contaminate the exposed edible in the areas after the kill step through to the sealing of the primary packaging. An EMP includes a visual check of clean equipment, real-time adenosine triphosphate (ATP) testing (explained later), and microbial testing for both indicator microorganisms and pathogens from swab samples of FCS and NFCS.

What if an edible does not have a kill step?

The development of edibles is rapidly growing, and edible manufacturers may be tempted to create an edible without a kill step. RTE edibles without a kill step are at high risk for pathogens. For example, in the development of a snack mix, the individual ingredients are procured from suppliers, and hazards are controlled by the suppliers with no further processing at the edible facility. The cannabis concentrate is warmed in a liquid used to coat one of the ingredients without enough heat to kill pathogens. The ingredients are blended and packaged for sale. An edible with no kill step requires verification of the control of hazards at the suppliers as detailed in Chapter 7, Preventive Controls. The scope of the EMP for an RTE edible without a kill step begins at batching or weighing up of ingredients where the ingredient is exposed to the environment, through any ingredient processing steps, blending, filling, and finishes with sealing of the primary package.

What if an edible is further processed after packaging?

The food industry has foods that are further processed to kill pathogens, and the facilities making these foods are not required to have an EMP. This makes sense when the biological hazard will be controlled after packaging. Examples include canned goods; bagged pickles meeting parameters of acidity, temperature, time, and benzoate concentration; high-pressure-processed foods; and frozen food baked or cooked in foodservice or by the consumer. In each example, the biological hazard is controlled with further processing of the product either within its packaging or after being removed from its packaging. I am not aware of an edible that is packaged and then further processed. When the time comes for edibles that are ready to bake, canned, or high pressure processed, the edible manufacturer will not need an EMP. Until then, every facility making an RTE edible must control the hazard of environmental pathogen with effective cleaning and sanitizing as a sanitation preventive control.

Hazard of environmental pathogen identified in the hazard analysis

The hazard of environmental pathogen is identified in the hazard analysis, in an edible safety plan for an RTE edible. The edible safety team or quality assurance manager is then free to design a facility-specific EMP to prevent cross-contamination of a pathogen from the environment into the product through effective sanitation and to use a combination of ATP, indicator microorganism, or pathogen testing for verification. FDA provides guidance but leaves the design of an EMP to the facility [1]. As with all FDA regulations for the food industry, food safety is built into the process and finished product testing is minimized. An EMP minimizes finished product testing and may not include finished product testing unless and until after a pathogen is discovered in the environment.

Pathogen testing of edibles

In contrast to food companies, cannabis companies are struggling to meet state code requirements for microbial testing of product and are burdened by the current system of sampling, hold and test, and release of product. By states dictating finished product testing requirements, cannabis companies may be missing hazards requiring control because the companies are so focused on meeting the state code, rather than an edible safety team analyzing the ingredients and process steps for hazards. At this point in history, cannabis companies must follow the state code for microbial testing and would be wise to integrate additional indicator tests and an EMP for the safety of their edibles.

The state code may dictate the testing of cannabis products for the foodborne bacterial pathogens *Salmonella*, Shiga-toxin producing *Escherichia coli* (STEC), or *Staphylococcus aureus*. *Aspergillus* is a pathogenic mold that may be required for testing. Testing for pathogens is expensive and time-consuming, and samples must be transported to a third-party laboratory. Testing for ATP and indicator microorganisms is comparably cheaper and quicker and can be performed on-site without the need to ship samples to a third-party laboratory.

Water activity (a_w)

Water activity is a growth factor for pathogens. In the same way that we humans require water, pathogens require water for survival and growth. Water activity is a measure of the energy of the available water needed for a pathogen to grow and not the same as percent moisture, a common test on an ingredient or product specification. Control of water activity of an edible is control of growth of pathogens. *Staphylococcus aureus*, the bacterial pathogen with the ability to grow at the lowest water activity among foodborne pathogens, will not grow below a water activity of 0.83. The goal to control the growth of pathogens is a water activity below 0.83. When dry ingredients or an edible has a water activity below 0.83, foodborne pathogens will not grow. There may be survival, but there will not be growth. Among foodborne pathogens, *Listeria monocytogenes* and *Salmonella* require a higher water activity for growth than *S. aureus* and are more easily controlled by lowering water activity. Most edibles have a low water activity that not only prevents the growth of pathogens but also prevents spoilage by yeast and mold. Shelf-stable edibles with a low water activity enjoy a long shelf life. Low-water-activity edibles include chocolates, hard confections, and gummies. Cannabis companies manufacturing bakery items may benefit from the routine testing of water activity on edibles. The water activity meter is small and reasonably inexpensive, being comparable to the size and cost of a pH meter. I recommend the monitoring of dried flower and edible finished product for water activity to verify stability through the shelf life.

The fallacy of finished product testing of cannabis-infused edibles with pathogen tests

The testing of edibles for hazards is too late. Building safety into the manufacturing and packaging of cannabis-infused edibles is the best place to apply resources rather than performing finished product testing of edibles. I understand testing of cannabis flower because there are so many products from flower; the variability of flower samples is inherently large and makes obtaining a representative sample very difficult. I support and recommend testing of concentrate for biological and chemical hazards; concentrate should inherently be a homogeneous sample, and test results from a small sample should be accurate for the full volume of the concentrate. I understand testing both concentrate and edibles for Δ^9-tetrahydrocannabinol (THC) and cannabidiol (CBD) potency for label verification. Beyond water activity and potency testing, I am not in favor of most testing of the edible finished product.

The requirement of Total Yeast and Mold Count (TYMC), pathogenic mold testing, and mycotoxin testing is a perfect example of the state dictating testing to the cannabis industry instead of cannabis companies writing their own preventive controls for the control of the hazard of mycotoxins. Good Agricultural Practices (GAPs), GMPs, and process preventive controls in growing, harvesting, and drying will control mold growth. In the hazard analysis for an edible, mycotoxins should be identified as a hazard requiring a preventive control for the concentrate as an ingredient, flour, tree nuts, peanuts, and potentially other ingredients. Every facility and every cannabis product are different, so it is best left to the cannabis company to identify and implement GMPs and preventive controls that will control mycotoxins. There is currently a heavy burden on the cannabis industry for testing to meet the state code for mycotoxins. There may be better use of funds in the prevention of mold growth to reduce the cost of finished product testing as the food industry has done.

Factors to consider for a representative sample

In addition to testing of edibles for hazards being too late in the process, there is the consideration of the accuracy of the results based on the edible sample being homogeneous and of a large enough sample size to be representative of the batch. The strongest argument to eliminate or reduce edible testing is that the statistical probability of finding a pathogen is extremely small and remote.

Homogeneity

Homogeneity is required for a representative sample. If the edible is a chocolate bar without particulates, soft confections, hard confections, beverages, capsules, or tablets, the sample is inherently homogeneous and sampling a representative sample is simple. Chocolate with particulates and baked goods such as chunky cookies may require aseptic blending before weighing the aliquot for microbial testing to obtain homogeneity.

Sample size

The sample size for microbial testing in the cannabis industry is unreasonably small, sometimes 1 g, meaning the sample is not representative of the whole. The quantitative methods common in both the food and cannabis industry include TYMC, Aerobic Plate Count (APC), Enterobacteriaceae (EB) count, coliform count (CC), and *E. coli* count. Quantitative methods in the food industry commonly start at 10, 11, or 25 g of a homogenized sample. The greater the sample size, the more accurate the counts will be and more reflective of the true count. Qualitative methods include hydrogen sulfide-producing thermoduric Enterobacteriaceae (HTEB), hygienic quality assay (HQA) (discussed later), and pathogen testing and commonly start with 25, 125, or 375 g of a food sample. Edibles meet the FDA definition of a Category I food and would be tested for *Salmonella* by collecting 60 samples of 25 g each for a total of 1500 g [2], if there is a chance of successfully finding *Salmonella* in the edible. Pathogens are assumed to be present in low numbers and starting with a large sample size increases the probability of the method detecting the pathogen. I am not advocating for the testing of 1500 g of samples, but 1 g of sample is certainly too small.

Probability of detecting a pathogen

Direct plating onto selective agar for quantitative testing: The manufacture and packaging of edibles is a harsh process for pathogens to survive. Pathogens that survive the removal of water and high heat of a kill step are injured cells. The methods that plate a diluted sample directly onto selective media, e.g., the method for *S. aureus*, do not allow for recovery of injury. Without repair of injury, there is no growth; without growth, there is no detection. However, those very same cells that are not detected are able to recover from injury under the right conditions and may cause an infection or intoxication in the body after consumption. Once the pathogen is in the body, there is warmth, water, and nutrients. Add time and the pathogen from the edible may cause illness.

Qualitative testing for pathogens: Qualitative testing for pathogens starts with sample enrichment to allow repair of injury and increase the cell number for detection. After enrichment, there are many method choices for detection. Qualitative methods result with either detection or no detection in a given product sample size or from an environmental sample. Instead of a count as from a quantitative method, most pathogen testing methods are qualitative and designed to detect pathogens in low numbers from large sample sizes. The probability of finding a pathogen when it is present is extremely low. Statistically, when 1% of product is contaminated with a pathogen, the number of samples needed to find the pathogen with a 95% confidence level of detection is 299 samples [3] typically from 25-g aliquots. This low probability of detection from 7475 g of an edible demonstrates how ridiculous pathogen testing is from 1 g of edible.

Finished product testing for pathogens is not recommended and may be a waste of resources based on the probability of detecting a pathogen, small sample sizes of edibles, and homogeneity of the edible. The solution is to build safety in by conducting a thorough hazard analysis of every ingredient and at every process step in a facility followed by implementation of GMPs and preventive controls to control the hazards.

Indicator tests

Indicator tests are an alphabet soup of acronyms used in both the food and cannabis industries and are useful tools to verify the sanitation of FCSs. The purpose of using an indicator test is to test for a related chemical, e.g., ATP, or microbial group without directly testing for a pathogen from an FCS. EMPs that require testing FCSs for pathogens are often the result of a misguided push from a customer. A positive pathogen result from an FCS sample implicates the entire batch of edible produced on that same day of sampling that was in contact with that FCS and until the surface was cleaned and sanitized. By finding the pathogen on the FCS, the pathogen is assumed to be in the product without directly testing the edible. With a positive test or high count from an indicator test instead of a pathogen test, the facility has no direct documentation of a pathogen in the edible and does learn that the conditions are favorable and have the potential for the presence of a pathogen. The facility can take the corrective action steps of cleaning and sanitizing followed by resampling.

Pathogen tests are very specific and may miss the pathogen if present in small numbers. In contrast, microbial indicator tests capture a larger group of microorganisms compared with a pathogen test. Indicator tests are done instead of pathogen testing because indicator tests are cheaper and faster than pathogen tests and can be done on-site. A robust EMP invests heavily in swabbing FCSs and testing swab samples for indicators rather than testing a small number of samples with expensive pathogen methods.

The data from indicator tests routinely done on FCSs can easily be plotted and analyzed. A baseline of results develops over time, which represents normal, clean conditions. A result that is a spike above the baseline indicates the need for intensive cleaning and sanitizing. Statistics can be applied in the analysis of data, upper limits can be set, and data can be reviewed for trends, changes, and seasonality.

Important points about indicator tests are as follows:

- *Indicator tests are verification activities for sanitation.*
- *The inspection of equipment and the environment being visually clean after sanitation is followed by swabbing for indicator tests.*
- *All indicator tests have advantages and disadvantages regarding turnaround time for results, supply cost, and technician training.*
- *For each indicator test, there are choices in laboratory equipment.*

Choices in methods abound, and a company should understand exactly what can be detected or measured from the method they choose.

Adenosine triphosphate bioluminescence

ATP bioluminescence is commonly used throughout the food industry. The real-time method of ATP testing is done on FCSs and in real time without leaving the site of the surface being tested. FCSs include equipment, bowls, trays, or utensils that come in direct contact with the edible. The ATP test is an indicator of the presence of any type of cell, including cannabis plant cells. The higher the ATP reading, the larger the number of cells, and therefore the surface has not been cleaned of food residue. Having a sterile surface is not the goal, so from most clean surfaces a low ATP reading is acceptable. ATP testing is best done after cleaning and before sanitizing because the sanitizer may interfere with the chemical reaction. ATP testing is done by swabbing a surface with a Q-tip-style swab and inserting the swab into a luminometer. The chemical reaction for bioluminescence is quick and completed without walking away from the site. If the reading is higher than baseline for the site or than the set goal, the corrective action is to reclean the FCSs followed by reanalysis for ATP. Typically, ATP testing is conducted by a quality assurance technician; it is only with the release of the equipment by quality assurance after a satisfactory ATP reading that manufacturing or packaging equipment can be used. There are several manufacturers of ATP swabs and luminometers. *ATP testing is highly recommended in an EMP.*

Staphylococcus aureus

S. aureus with a nickname of "staph" is associated with employee hygiene and employee handling of food. Almost half of us humans carry staph in our nasal passage and on our skin. Think of an employee sneezing over edibles and spreading staph; that is gross, and the edibles should be discarded. Employees must follow GMPs to prevent the cross-contamination of staph from the employee to the edible. Staph causes foodborne illness after the cells grow in a temperature-abused food and produce an emetic toxin. The concentration of cells in the food must be greater than 100,000 colony-forming units (CFUs) of staph per gram. That's a lot of growth and a lot of cells! One potential example I see of getting enough cells in an edible is if a hand-kneaded dough or manually stirred batter for a bakery item was left out overnight and then baked the next day. Don't do that.

It is not the staph cells that cause a foodborne illness, it is the toxin produced by the cells as they grow. No growth means no toxin, which means no illness. Only direct testing for toxin will determine the presence of toxin, which is time-consuming, requires a high level of expertise at a third-party laboratory, and is costly. The testing for staph in an edible is an indicator test for the presence of toxin and a very poor indicator; cells may have died or become injured and may become undetectable while the toxin produced by the cells is active. Edibles, dried cannabis, or dried hemp with a water activity below 0.83 will not have enough available water for the cells to grow. Monitoring of water activity is an excellent method in the implementation of a process preventive control for drying and finished product testing. Controlling water activity will control the growth of staph. By preventing the growth of staph, the production of toxin is not possible.

Companies concerned about staph in ingredients may want to implement a supply chain preventive control for the control of staph at their supplier. See the section Supply chain preventive controls in Chapter 7, Preventive Controls. *S. aureus* testing is not routinely done in the food industry, and finished product testing of edibles for staph is not recommended by this author. *S. aureus testing is not recommended in an EMP.*

Total Yeast and Mold Count

Mold is naturally present in the growing operations of cannabis and hemp. Weather conditions for outdoor growth of cannabis or hemp and harvesting damage the plant and create the conditions for mold growth. Control of drying operations is critical to control mold growth and the potential production of mycotoxins, which are poisons produced by mold. The discussion in the *S. aureus* section applies here for toxin production and monitoring of water activity as a process preventive control. In most mold growth, it is not the mold cell that causes illness but the mycotoxin produced during mold growth.

TYMC quantifies both mold and yeast because of the different colony morphology in the growth medium; mold are fuzzy and yeast are not. In general, yeast are not a concern for foodborne illness but may shorten the shelf life of an edible with spoilage. The state code may set a maximum TYMC of 10,000 CFUs per gram of sample. There are some limited remediation steps for plant material with a TYMC above the limit, e.g., treatment with ozone, chlorine dioxide, or heat. Most cannabis products, i.e., concentrate or edibles, over the limit historically have been discarded and never entered the market. There is a concern that if plant material with mycotoxin is used to make a concentrate then the concentrate will have a high content of mycotoxin. It makes sense for edible manufacturers to review the results for mycotoxins in concentrate as a supply chain preventive control. Verification of the supply chain preventive control should include record review of parameters in the drying operation for the harvested cannabis or hemp.

Most mold detected in the TYMC are nonpathogenic. Some states do require the direct testing of pathogenic mold such as *Aspergillus* or of specific mycotoxins; these methods are lengthy, labor-intensive, and costly. In comparison, the indicator method of TYMC is also time-consuming and takes a minimum of 4 days for aerobic growth at 25°C, i.e., close to room temperature. The TYMC method is cheap, is easy, and can be done on-site at a cannabis company, in addition to required testing done at the cannabis laboratory.

TYMC is an indicator method that is used widely throughout the food industry for air quality, ingredient specifications, and finished product. TYMC is a useful tool to monitor the seasonality of air quality, with higher mold counts expected at certain times of the year. TYMC is an indicator of product spoilage but not a good indicator of food safety. As an indicator method, TYMC has the potential to monitor general sanitary conditions of FCSs in an EMP. However, TYMC does not indicate the potential of a pathogen on an FCS. *TYMC testing is not recommended in an EMP.*

Aerobic Plate Count

APC is used extensively throughout the food industry and may be required by the state code for cannabis products. A better use of the APC than finished product testing is testing FCSs as part of an EMP. Typically, environmental samples from NFCSs, such as floors and drains, are not tested by APC because the APC would be too high due to the normal presence of bacteria on those sites. The APC of FCSs is most valuable with a history of data from a given site; a facility can determine when a high count indicates poor sanitation on an FCS in comparison to a baseline APC. An APC of <1000 CFUs per clean FCS sample site is in the ballpark for indicating a clean surface. The laboratory on-site at a cannabis company can test for APC; the method for APC is simple, cheap, and fast. APC results are available after 48 hours of aerobic incubation of plates at 35°C. The bacteria that grow in APC method have similar growth characteristics as many of the pathogens responsible for foodborne illness, making the APC a good indicator test.

APC is an indicator test for the general number of bacteria on a surface. Many bacteria present in the sample will grow and be detected, but not all. *S. aureus* will grow under the APC conditions and it is difficult to see separately from the other bacteria in the sample. EB, discussed later, will grow. Despite the synonyms of Total Plate Count or Total Aerobic Microbial Count for the method's name, this count is not a total count. Bacteria with greater nutritional requirements, with slow growth, or that are unable to grow in the presence of oxygen, i.e., anaerobes, will not be detected by APC. The sample is incubated for 48 hours, but results may not be reported from a third-party laboratory until 3 or 4 days after the sample was taken. When a high count is reported, the corrective action is typically to resample the site after an intense cleaning of the manufacturing or packaging equipment. Any disadvantages withstanding, *APC testing is highly recommended in an EMP for FCSs.*

One of the following three methods is highly recommended in an EMP: EB count, CC, or *E. coli* count. It is not appropriate and would incur unnecessary expense to conduct both the EB count with either the CC or *E. coli* (EC) count. The CC/EC Petrifilm method provides both a CC and *E. coli* count. Note that the sample is set up once, not twice, onto one medium for the CC/EC Petrifilm method, so a client should not be overcharged by a laboratory for two separate Petrifilm methods. After 24 hours the CC is recorded. The medium is reincubated for an additional 24 hours after which the *E. coli* count is recorded.

Enterobacteriaceae count

When I think of the family EB, I think of a big family reunion. The family members share a handful of similarities but there is also great diversity in the family. EB (Fig. 8.1) is a family of bacteria including the genus *Escherichia*. The method for EB count is simple, cheap, and fast, with incubation stopped at 24 hours. The laboratory on-site at a cannabis company can test for EB count. Members of the family EB are commonly found in nature and some but not all are pathogens. Clean surfaces should yield an EB count of <10 CFUs per sample site; this means EB was undetectable. When a laboratory reports any EB count of 10 or more from an FCS, there is no indication of pathogen or nonpathogen. A completely different method is required to detect pathogens. *Reporting of EB counts of >10 CFUs per sample site from clean food contact surfaces is an indicator of poor sanitation and requires corrective action. Of the three methods, EB count is highly recommended in an EMP for FCSs.*

Coliform count

The coliform group is a group of bacteria from the family of EB, including generic *E. coli*. Members of the coliforms are commonly found in nature and considered nonpathogenic. The coliforms share some characteristics of the family EB, with all members of EB like family members at a family reunion all descending from the same pair of grandfather and grandmother. Coliforms share additional characteristics with each other that make them a unique group, like all the cousins who have tattoos. Notice the term "coliform" is not italicized or capitalized; the term is not a proper name like the names given in italics. The on-site laboratory at a cannabis company can test for CC; the method for CC is simple, cheap, and fast. When a laboratory reports a CC other than <10 CFUs per site, coliforms were detected from the

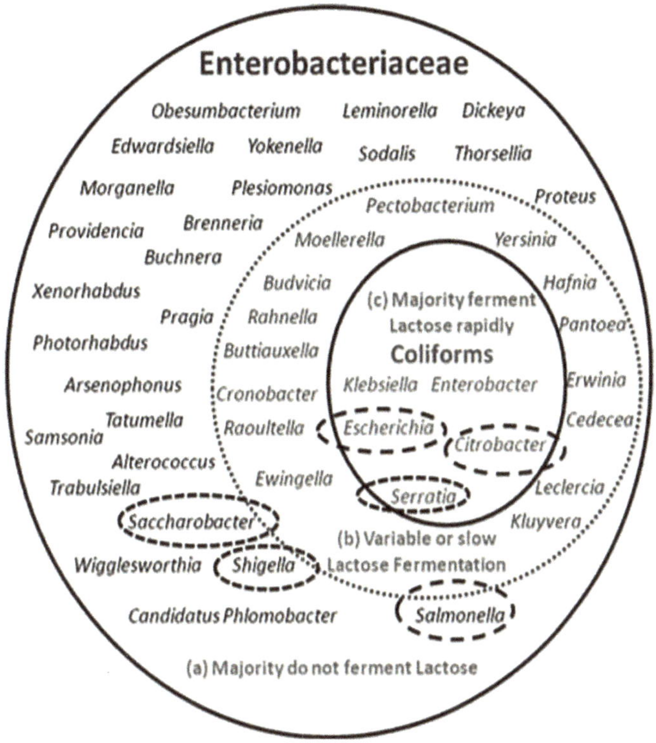

FIGURE 8.1

The family of Enterobacteriaceae includes pathogens and nonpathogens. Coliforms are a group of bacteria within the family, mostly nonpathogenic, and found in nature. Pathogenic *Escherichia coli* is not characterized as a coliform and is of fecal origin [4]. *Used with permission from the International Life Sciences Institute.*

sample but not necessarily *E. coli* because *E. coli* is just one member of the coliform group. *Reporting of coliform counts of >10 CFUs per sample site from clean FCSs is an indicator of poor sanitation and requires corrective action.* Because coliforms are commonly found in nature, there is not a correlation to fecal contamination as is the case with *E. coli*.

E. coli count

Escherichia is one genus in the large family of EB and one member of the coliform group. *E. coli* is one species in the *Escherichia* genus and is the major portion of feces from all animals, including you! Generic *E. coli* is nonpathogenic; pathogenic *E. coli* is discussed separately. When a laboratory reports an *E. coli* count other than <10 CFUs per site, *E. coli* was detected. The original source of that *E. coli* was fecal matter. However, the fecal matter is probably not present, and the original contamination of fecal matter is long gone. *E. coli* may come into a facility through rodents, sewer water backup, drain aerosol, bird droppings on the roof, raw ingredients, or people. Spices and seasoning blends, if not irradiated, are notorious for being a source of *E. coli*, as is fresh produce or any raw animal product. Once inside the facility, the contamination may travel via roof leaks, dripping condensate, water spray, or water drainage. Many human pathogens are transmitted through fecal contamination. If nonpathogenic generic *E. coli* is present, there is a probability that a pathogen of fecal origin could be present. Finding *E. coli* on clean FCSs is not acceptable and requires rigorous cleaning and sanitizing with additional sampling of the site and related areas surrounding the site as part of corrective action. Once established in the processing environment, *E. coli is an indicator of fecal contamination and of very poor sanitation. An E. coli Count of >10 CFUs per sample site requires corrective action.*

The laboratory on-site at a cannabis company can test for generic *E. coli* because generic *E. coli* is nonpathogenic. The method for generic *E. coli* is simple, cheap, and fast. Some state code requires the testing of cannabis product for STECs. Testing for generic *E. coli* is not the same as testing for STECs! When a cannabis laboratory tests for STECs from a sample of cannabis-infused edible, the laboratory uses a completely different method that is more costly, requires a high level of laboratory competence, and takes more time than the method for generic *E. coli*. Even though STECs are *E. coli*, not all *E. coli* are STECs. The STEC method specifically detects pathogenic *E. coli* and not generic *E. coli*.

Are generic *E. coli* and coliforms present on the plant material of unprocessed cannabis? Yes! I expect coliforms to be found on cannabis grown outdoor or indoor because it is completely normal to find coliforms on plants in nature. I expect generic *E. coli* to be found on hemp and cannabis grown outdoors from fecal contamination by birds and rodents. Growing operations must achieve containment of *E. coli* and coliforms through facility design, air handling, control of water, traffic patterns, dedicated tools and supplies, dedicated personnel, and personal protective equipment. With control, *E. coli* and coliforms will stay in the growing operation and not cross-contaminate adjacent processing areas. Containment of *E. coli* and coliforms is best achieved in dedicated facilities without processing steps beyond drying and trimming.

Kornacki Microbiology Solutions, Inc.

Two indicator tests have been developed by my colleague, Jeff Kornacki: one is an indicator for *Salmonella* and the other for *Listeria* [5,6]. As consultants in the food industry, Dr. Kornacki and I sample FCSs in recall investigations or risk assessment projects seeking to find the source of a pathogen. If the pathogen of concern is *Salmonella*, the samples are tested by the HTEB method. If the pathogen of concern is *Listeria*, the samples are tested by the HQA method. Both methods indicate the potential presence of the pathogen without directly testing for the pathogen. Environmental samples from NFCSs can be directly tested for the pathogen of concern if

FCSs and edibles are not implicated from the sample site. If we instructed the laboratory to run a pathogen test on a sample from an FCS and the pathogen was detected, there would be regulatory implications. When a pathogen is found on an FCS, FDA believes the pathogen is in the product and requires a recall or an extension of an existing recall if more product is implicated. For these reasons, the HTEB and HQA methods were developed as indicator tests for the pathogens *Salmonella* and *L. monocytogenes*. As indicator tests, there is a high rate of false-positive results, as the methods detect a large group of bacteria. A positive indicator test result shows the potential presence of a pathogen.

Hydrogen sulfide-producing thermoduric Enterobacteriaceae for Salmonella*-like bacteria*

Salmonella is a member of the family EB, along with coliforms and *E. coli*. Unlike the latter, all *Salmonella* are pathogens. *Salmonella* originates from fecal contamination from animals with an infection and can make its way into an edible facility and harbor in niches in the environment. Through cross-contamination from the environment, *Salmonella* may be transferred to an FCS and ultimately into the edible. *Salmonella* is associated with dry environments and RTE edibles with low water activity. This makes *Salmonella* an environmental pathogen of concern in most cannabis-infused edibles. A positive HTEB result indicates the potential presence of *Salmonella* and requires intensive corrective action. *HTEB testing of FCSs is highly recommended in an EMP.*

Hygienic quality assay for Listeria*-like bacteria*

Listeria is ubiquitous in nature and within edible facilities. *Listeria* is associated with wet environments and RTE edibles with high water activity, such as beverages. I even heard of ice cream being made as an edible, and *L. monocytogenes* is a pathogen of concern in ice cream. A positive HQA result indicates the potential presence of *L. monocytogenes* and requires intensive corrective action. *HQA testing of FCSs is highly recommended in an EMP.*

Pathogen testing for environmental monitoring

Identify hygienic zones

The food industry employs a hygienic zoning system that defines zone 1 as FCS and zones 2–4 as NFCSs. Zone 1 sites and FCS sites are analogous and interchangeable in speech and writing. Samples from zone 1 sites are often tested by indicator tests, but NFCS samples from zones 2–4 are tested for an environmental pathogen of concern. Zone 2 sites are adjacent to and associated with FCSs but are actually NFCSs. Zone 3 surfaces typically include doorways and thresholds to the processing and packaging rooms and the room's ceiling, walls, floor, and drains. Zone 4 sites are outside the processing and packaging rooms. As many cannabis manufacturing areas are quite small, walls and room structures may be classified as Zone 2 rather than

Zone 3. Ceiling structures and overhead pipes directly above open product are classified as Zone 1 as dust or condensate will drop down onto the product.

Pathogens enter the facility with wheeled equipment, ingredients including water, people, air, and rainwater. Pathogens travel throughout the facility with wheeled equipment, ingredients, people, air, uncontrolled water, and tools and utensils. It is normal for pathogens to enter a facility. Think of employees who are ill and carry in the pathogen. Think of raw hemp grown outdoors and received at an extraction facility. Containment of a pathogen in the farthest areas away from the exposed edibles is critical, and these areas are labeled as zone 4 as being the farthest away. Every facility defines the zones of their unique facility and the sample sites within those zones.

On a facility blueprint or schematic of the rooms, first label zone 4 because this zone is the easiest to identify. Zone 4 is outside the processing and packaging areas and typically includes Receiving, Ingredient Storage, Batching (except edibles without a kill step), Edible Storage for edibles sealed in packaging, and Shipping. The ideal facility has the separation of Receiving and Shipping to control environmental pathogens. A limited number of samples will be taken from surfaces outside the processing and packaging rooms to verify control of pathogens making their way into the room. It is appropriate to sample floors from hallways outside the processing and packaging areas and sites along typical traffic patterns of employees and wheeled equipment. It is critical to control pathogens in some zone 4 areas to prevent the pathogen from entering the processing and packaging rooms. Some zone 4 sites on the raw side may logically be positive for the pathogen, and the pathogen testing would not be helpful in those areas. For example, if the facility is receiving raw hemp, it is not possible to prevent a pathogen from entering the facility at Receiving. What good would it do to confirm the presence of a pathogen at Receiving? A facility is obligated to conduct corrective action after a positive pathogen result. Repeated caution must be given when sampling sites, starting with zone 4, to understand the implication of positive results from each site and know the corrective action that would be applied before taking the environmental sample.

The work of the facility is containment of the environmental pathogen and prevention of the pathogen from entering the processing and packaging rooms. If the pathogen is outside the post-kill-step area, it can enter the room. If the pathogen is found in the room, it can be transferred to the surfaces of equipment. If the pathogen is on a piece of equipment, it can be transferred to an FCS. If the pathogen is on an FCS, it can be transferred onto the edible. By sampling and testing environmental sites, a facility draws a picture of the control of environmental pathogens or the need for corrective action.

I next locate the kill step on the facility schematic, if there is a kill step. Processing steps before the kill step and into the kill step are on the raw side and often labeled zone 4. Imagine a push-through dishwasher where dirty dishes go into the dishwasher on one side and cleaned and sanitized dishes come out on the other side. Most of the kill steps in the food industry follow this process of raw product in one side of the kill step and finished product out another side. Examples include pasteurizers, roasting ovens, and belt-fed baking ovens. In contrast, a cannabis kitchen has open kettles for cooking and bakery ovens for baking. The area of the kill step itself is not labeled zone 4 but is labeled zone 3. A bakery oven has raw product enter and finished product exit, so the area is zone 3 because of the potential for cross-contamination after baking. Every processing and packaging room or area the finished product enters after the kill step and where the edible is exposed to the environment includes zone 3 sites of the thresholds to the room, ceiling, walls, floor, and drains. If a rack of work-in-process or nonpackaged edibles is moved into a cooler or freezer, the path to the cooler or freezer and the entire cooler or freezer is zone 3 not zone 4 because of the potential cross-contamination from the environment. In small facilities, traditional zone 4 sites outside the processing or packaging area may be zone 3 sites because of the proximity of the area to the edible.

Every FCS the edible touches and the associated NFCS of equipment are zones 1 and 2, respectively. Gloved hands that manually move the edible are zone 1 FCSs. Utensils used for stirring or transferring the edible are zone 1 sites. Every structure above the exposed edibles is zone 1, as a pathogen could be carried down to the product on air currents, a dust particle, or water droplet. The motor assembly of a Hobart mixer above the bowl is zone 1. Overhead pipes and cables are zone 1. An air duct that blows over exposed edible is zone 1. The inside housing of a cooling tunnel is zone 1. In small cannabis kitchens, the traditional zone 3 sites of threshold, ceiling, walls, and drains may be zone 2 sites because of the proximity of the area to the edible. For example, sites in my home kitchen would be zone 2 without zone 3 sites because of the proximity of the sites to the edible. The half bath next to the kitchen would be zone 2; try not to imagine the possibility for cross-contamination there. The landing to the garage would be zone 3. Control of environmental pathogens focuses on employee hygiene, use of personal protective equipment, and effective cleaning and sanitizing of the entire processing and packaging rooms from ceiling to drains. Model food safety plans are available from the Food Safety Preventive Control Alliance to illustrate zoning (Fig. 8.2).

Determine the pathogen of concern in the environment

Do not confuse the pathogen of concern to control through processing of an edible with a pathogen of concern in the processing environment. The pathogen of concern in an edible is dependent on the ingredients, the control with a kill step, and properties of the edible such as water activity. An EMP is all about controlling the hazard of an environmental pathogen cross-contaminating the edible from an FCS or NFCS. The pathogen of concern in the environment depends on whether the post-kill steps are in an area where the environment is dry or wet, the proximity of environmental sites to the edible, and the methods of cleaning and sanitizing of utensils, equipment, and rooms.

In general, *Salmonella* is the environmental pathogen of concern in edibles in which the post-kill steps occur in a dry environment. Examples include bakery items after the oven, hard confections and gummies after the cooking of syrup, and most

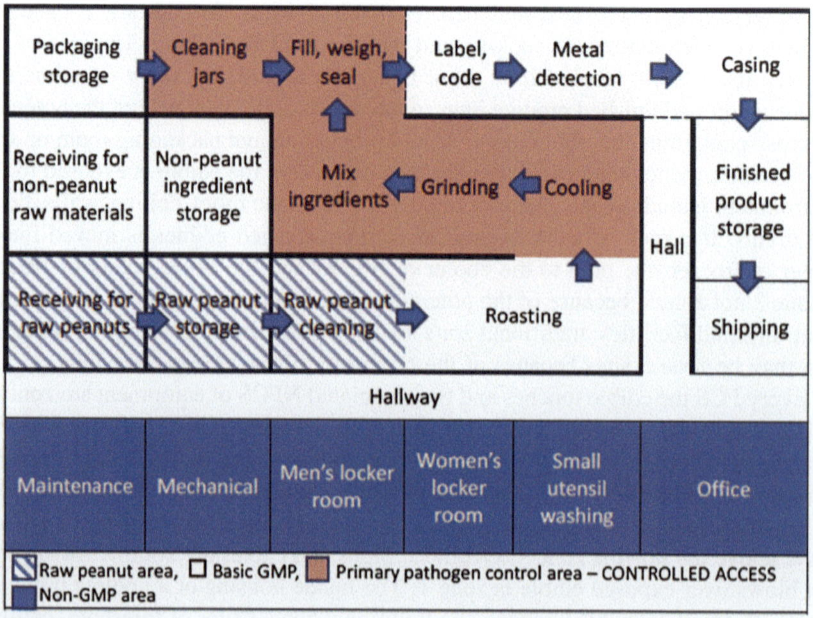

FIGURE 8.2

Blueprint schematic of a peanut roasting and peanut butter facility used for training only [7]. Sample sites outside the primary pathogen control area are zone 4 samples; note the figure key for raw peanut area, basic Good Manufacturing Practice (GMP), and non-GMP area. The kill step is roasting. Zone 1 sites are food contact surfaces of equipment starting with conveyance of roasted peanuts to cooling and are located in the primary pathogen control area. Thresholds, ceiling, walls, and drains in the primary pathogen control area are zone 3 sample sites. Zone 2 sites are dependent on the size of the rooms and the proximity of equipment to the exposed product.

steps in chocolate making. Because there is no kill step in chocolate making, unless the chocolatier starts with roasting of cacao beans, the EMP includes all areas where the ingredients and chocolate are exposed to utensils and the environment. *Salmonella* is known to persist in a dry environment and spring to life when water is introduced. There are dry processing equipment and environments in the food industry in which it is riskier to clean and sanitize in the traditional wet manner because of the introduction of water into the system than to continue processing without the benefit of traditional wet cleaning and sanitizing. Alternative dry clean procedures are appropriate, including the physical removal of residue from surfaces, use of alcohol wipes, and use of steam, which evaporates and leaves a dry surface.

In general, *Salmonella* is the pathogen of concern in edibles with low water activity, but only if the post-kill step area is dry. *L. monocytogenes* is the environmental pathogen of concern in edibles in which the post-kill steps occur in a wet environment. Examples of edibles made in a wet environment include beverages and ice cream. In wet environments, it is typical to find water on the floor during production and cleaning and sanitizing of all utensils, equipment, and rooms is conducted daily with traditional wet cleaning and sanitizing steps. Owing to many cannabis kitchens being small and despite most edibles being a low-water-activity product, the post-kill step area in a cannabis kitchen is often wet or in proximity to a preparation area or sink area that is wet. In a cannabis kitchen with wet areas, the pathogen of concern from the environment is *L. monocytogenes*.

The combination of making low-water-activity edibles in a wet cannabis kitchen means that both *Salmonella* and *Listeria* are environmental pathogens of concern. With the same concept, a drain is always wet and a potential harborage site of *Listeria*, even in dry environments where the environmental pathogen of concern is *Salmonella*. Drain samples from a dry environment should be tested for both *Listeria* and *Salmonella*. If a pathogen is present in the environment and finds its way to a drain, there is water, food, and time for growth. All things go to the drain (in a takeoff from the movie, *Finding Nemo*, where all things lead to the ocean), so a drain is a good sample site that is representative of the area. Drain samples are highly recommended in EMPs from both wet and dry environments because drains are wet even in dry environments and may be a source of pathogens.

Listeria monocytogenes

L. monocytogenes is a pathogen that causes infection after being ingested. Immunocompromised patients who consume an edible contaminated with *L. monocytogenes* are more likely to get a *Listeria* infection because of the patient's weakened immune system rather than a healthy adult with a healthy immune system. Immunocompromised persons include the very young, the very old, pregnant women and their fetus, people under medical treatment such as chemotherapy or prescription drugs, and people with medical conditions. Patients on medical marijuana are at risk for listeriosis. In a typical *L. monocytogenes* outbreak the mortality rate is around 20%; I have seen as high as 40% reported. It is not known how many *L. monocytogenes* cells need to be consumed to cause illness, so the goal is to have undetectable *L. monocytogenes* in edibles. FDA has zero tolerance for the presence of both *Salmonella* and *L. monocytogenes* in foods, meaning any detection of the pathogen in the food or on an FCS makes the food adulterated resulting in a recall of product in the market.

There are choices for *Listeria* testing, and the method chosen should be validated at the cannabis laboratory. With the choice of method comes the choice of testing for either *Listeria* species (spp.) or *L. monocytogenes*. There are 17 species of *Listeria* to detect in a *Listeria* spp. method; *L. monocytogenes* is the species that is pathogenic. *L. monocytogenes* can be found in environmental samples, along with other species of *Listeria*. For example, *Listeria innocua* outcompetes *L. monocytogenes*, with *L. monocytogenes* being present in relatively small numbers and making detection difficult. My colleague and I had a drain sample from a client's facility test positive for both *L. monocytogenes* and *L. innocua*. We agreed that the drain must have been heavily contaminated with *L. monocytogenes* for the method to detect both species.

The other method of choice is very specific for only the pathogen *L. monocytogenes* and not the other *Listeria* species. Areas that potentially harbor *L. monocytogenes* may be reported as negative, with the very narrow focus of detection and other *Listeria* species outcompeting the pathogen. The choice of *Listeria* spp. method is often better than the choice of *L. monocytogenes* method. *Listeria* spp. methods serve as an indicator of the presence of *L. monocytogenes* and corrective action should follow as if the pathogen had been detected. The FDA's current thinking is that a positive test result for *Listeria* spp. should be considered positive for the presence of *L. monocytogenes*, even without confirmation of the species. For any method chosen and before a site is sampled, the company needs to think through the implication on product and write the steps for corrective action when the pathogen is detected.

Listeria is ubiquitous. I love that term. *Listeria* can be found on a multitude of sites from different food processing environments; always sample the drains. *Listeria* is in the edible environment. I don't need a peer-reviewed, published research study to tell me that. I will be more than happy to do that project if someone wants to pay me to conduct a project of sampling and testing cannabis kitchens for *Listeria*! Leaders among cannabis-infused edible manufacturers will implement an EMP without having a regulatory requirement, and customers will drive the requirement for an EMP before regulation does. If this is new information, don't fret. The food industry and all its associated service providers of software, equipment manufacturers, and chemical companies have your back. It is only a matter of time before cannabis-infused edible manufacturers will be required to implement an EMP for *Salmonella* and *L. monocytogenes*. There are more resources on EMPs than you will ever use.

The food industry would be shocked!

The cannabis industry relies heavily on testing of flower, concentrate, and edibles. The cannabis industry lives in a land where flower, concentrate, and edibles are held, tested, and released if they pass testing. I have some disagreements with flower testing. I think monitoring of GAPs can substitute for most of the flower testing, as the food industry has held farmers accountable for the safety of produce and raw agricultural commodities. I absolutely agree with the testing of concentrate as a supply chain preventive control for an ingredient in an edible. That leaves the testing of edibles, and the food industry may be shocked by the requirement to hold all edibles until laboratory results are released. The historical precedence of the food industry to monitor processing steps goes back to the 1950s and 1960s when the food industry relied heavily on finished product testing, which is where the cannabis industry is now. NASA was preparing for the first manned mission to the moon [8]. NASA required a higher level of confidence that the food destined for space was safe for consumption than was the current level in the food industry. Imagine the disaster of fever, diarrhea, or vomiting in space from foodborne illness. The work that followed to improve food safety revolutionized the food industry for over 50 years all the way through to our current FDA rules in the Food Safety Modernization Act [9].

The food industry relies heavily on verification of the safety of ingredients and monitoring of process steps to build safety into the product and minimize finished product testing. It is possible for an RTE food facility to never test their finished product and the product to be safe. This may be shocking to the cannabis industry. For example, Romaine lettuce cannot be tested for pathogenic *E. coli* because the shelf life of the lettuce is only 3 weeks. It is not possible to hold the lettuce until laboratory results are released. I realize Romaine is not the best example owing to the many outbreaks and cases of foodborne illness caused by Romaine. The farming and produce industries are working diligently to build safety in and protect consumers. When an edible safety team completes a hazard analysis and identifies preventive controls for each hazard, the hazard should be controlled without finished product testing. Wouldn't it be wonderful to live in a land where edibles are shipped to a dispensary immediately after packaging? Dosing will still require some frequency of testing of the edible, but the goal is to build safety in and minimize finished product testing.

Initiation of an environmental monitoring program

Companies should take strong measures to separate drying and trimming operations to prevent cross-contamination. A best practice (perhaps a requirement) is to have dedicated personnel who place the harvested cannabis in the drying room separate from personnel who monitor and collect the dried cannabis and separate from the trim team. The ideal facility design would have the harvested cannabis enter the drying room through a door on the growing side. The drying room floor would be vacuumed after the room was filled. A second door on the trim room side would allow personnel to enter the drying room to monitor and collect the dried cannabis without walking through the growing side. After the dried cannabis is moved to the trim room and the drying room is emptied, the entire room would be cleaned, sanitized, and dried before filling the drying room, with the next crop of harvested cannabis.

Clean break

The documented procedure of cleaning and sanitizing a room or equipment is called a "clean break" and is critical to control cross-contamination and limit the scope of a recall. Effective cleaning and sanitizing procedures will remove and kill pathogens on equipment and in rooms. If a company were to initiate a recall of cannabis or an edible due to a biological hazard, a lot is bookended and defined with the clean break before the product was made and the clean break after the product was made. In the food industry, an entire line of equipment must be completely shut down, cleaned, and sanitized to effectively create a clean break. Two clean breaks define a lot in the eyes of FDA.

Use of dedicated HEPA-filtered vacuums

An indoor growing operation is a wet environment, and employees can sweep to collect debris and spray down the floor as needed. The use of water and strong air movement creates an environment where all plants are exposed to all contamination within the room. The drying room is a dry environment. The use of brooms is not recommended owing to the generation of air currents carrying dust, debris, and contamination. When a broom is used in a drying room filled with drying cannabis, whatever is on the floor is now on the plant, and the goal is to protect the plant from cross-contamination during drying. Instead, the recommendation is to vacuum debris using a HEPA-filtered vacuum dedicated to the drying room. Because the vacuum releases a strong air current, whatever is exhausted from the vacuum is now on the plants. Maintenance of the dedicated HEPA-filtered vacuum is critical for the control of cross-contamination. A second HEPA-filtered vacuum should be dedicated to the trim room, and additional HEPA-filtered vacuums should be placed in any other dry environments with exposed product and not moved from room to room.

Sanitation preventive controls

Sanitation preventive controls go hand in hand with EMPs. To initiate an EMP shore up the sanitation standard operating procedures (SSOPs) with written procedures, documentation, and training. After all, the EMP is verifying the effectiveness of the SSOPs. Dedicate the use of and implement a color code for cleaning tools, maintenance tools, and utensils to prevent cross-contamination. Invest in new hoses when old hoses fray.

Create a master site list

The sampling plan includes a master site list for the facility that details the sites to be swabbed and notes their zone. FCS sites should include the most difficult to reach and clean parts of equipment. A small edible facility potentially may have a master site list of 20 sites in total. On the other hand, I was in a large food facility that collected over 200 swab samples weekly for testing. The sampling plan is unique to the facility, edible, and equipment. The design of the sampling plan focuses on verifying that the sanitation procedures are preventing cross-contamination of the edible.

The frequency of swabbing for environmental monitoring is determined in the evaluation of the variables of equipment design, room and facility design, foot traffic, wheeled equipment traffic, air, historical results, and other factors. One FDA recommendation is to swab five FCS sites per line per week [1]. How a "line" is defined is difficult, with the onus on the edible safety team. The master site list is used to determine which sites are swabbed on a weekly basis. Sites from the zone 1 list are swabbed weekly, and in general, the completion of the rotation of zone 1 sites is monthly. In other words, all zone 1 sites are swabbed monthly. To the other extreme, the completion of the list of all zone 4 sites may be quarterly, and a facility may choose to swab zone 4 sites monthly and not weekly. Completing the list of zone 2 and zone 3 sites lies between the two cases for zones 1 and 4, with zone 2 and zone 3 sites swabbed weekly, biweekly, or monthly with the rotation completed monthly, within 2 months or quarterly. Weekly swabbing may or may not include zone 2 and zone 3 sites. There are too many variables from facility to facility to describe a one-size-fits-all program.

Most EMP sampling plans build in an element of randomization. Some companies use software to pick the sites for sampling. In general, it is better to identify a long list of sites and randomize the sampling than to sample a short list of sites over and over. The same is true for the day of sampling and time of day. The more a facility can mix up the days and times of sampling, the greater the probability of finding an environmental pathogen if present. For both indicator tests and pathogen tests, a facility develops the picture of control of pathogens and effectiveness of sanitation, and the EMP will adjust for the location of sites and the number and frequency of sites sampled.

Seek and destroy

The military uses the term of *seek and destroy* when combating our enemies. Great effort is placed in surveillance of the enemy, and soldiers are provided with the weapons to destroy the enemy when needed. The concept of seek and destroy carries over to the food and cannabis industries for the control of our enemy, pathogens. An EMP is a surveillance program designed to find the enemy. If all environmental sample sites achieve low or no counts and pathogens are never detected, environmental monitoring is not successful. Environmental monitoring starts with the premise that pathogens are entering the facility, and an occasional positive result for a pathogen from an NFCS is expected by and acceptable to inspectors. I am concerned about a facility that never has a positive pathogen result, because the lack of positives tells me there is a weakness somewhere in the EMP.

What to do when a site confirms positive for a pathogen

FDA guidance offers steps to follow when a positive pathogen result is reported from an FCS (Fig. 8.3) or NFCS (Fig. 8.4) [1]. As guidance, this is FDA's current thinking on the topic, and a company is not required but is recommended to follow these steps. This guidance addresses the control of *L. monocytogenes* in RTE food and can be applied in the control of *Salmonella* and RTE edibles, or companies are encouraged to design reasonable corrective action steps that work for their facility. If a company chooses a different path from the FDA guidance, the edible safety team must be prepared to justify their decisions for corrective action. The edible safety team should talk through and design the corrective action steps for both FCSs and NFCSs before a sample is sent to the laboratory. Do not wait until a positive pathogen result from an NFCS is reported to decide the next step. Corrective action must be done after a positive pathogen result.

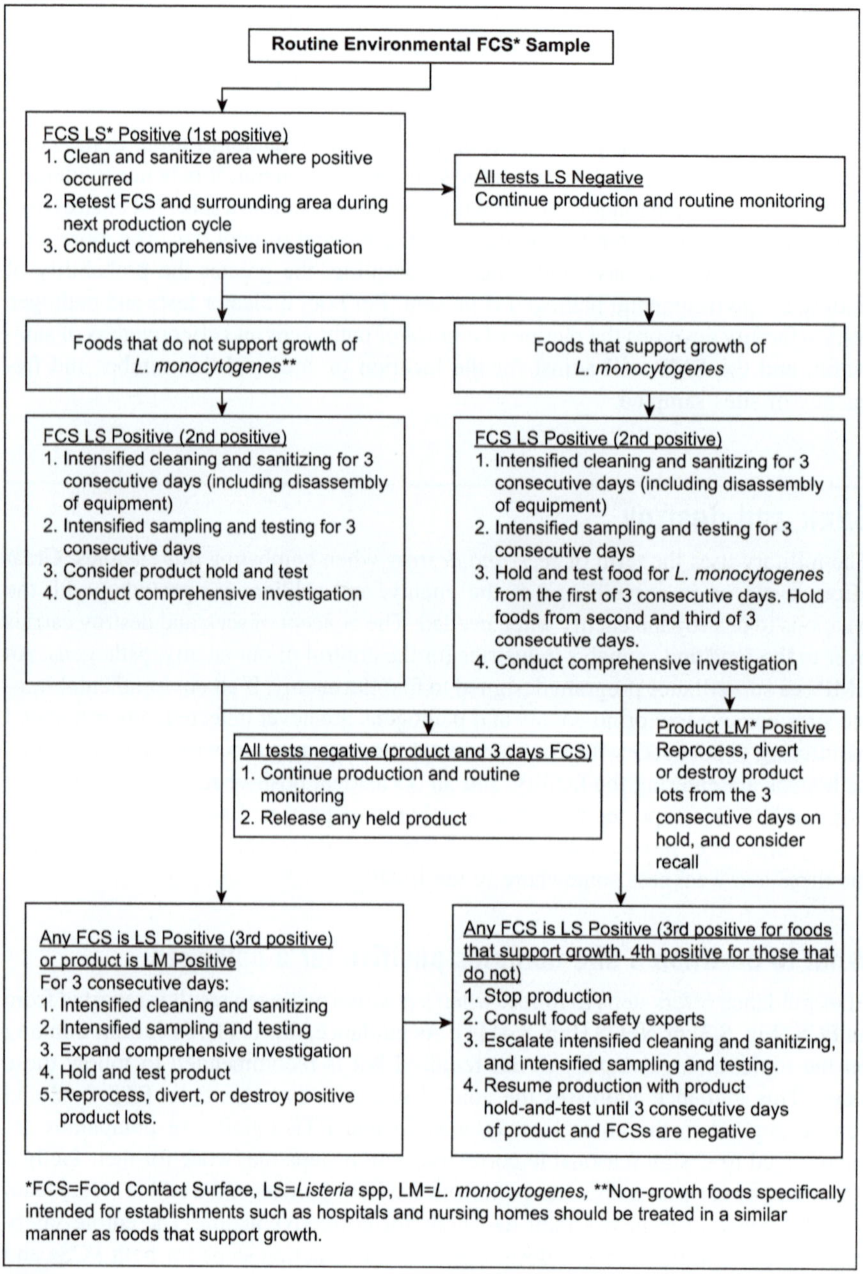

FIGURE 8.3

Routine swabbing of food contact surfaces. *FCS*, food contact surface; *LM*, *Listeria monocytogenes*; *LS*, *Listeria* spp. [1].

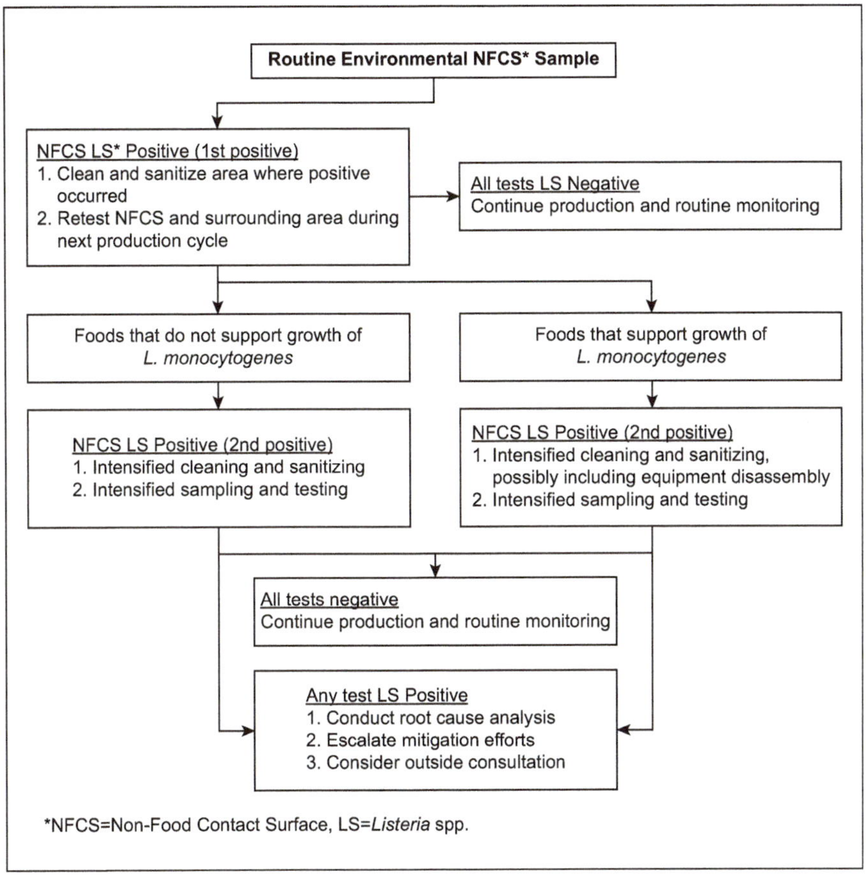

FIGURE 8.4

Routine swabbing of non-food contact surfaces. *LS*, *Listeria* spp.; *NFCS*, non–food contact surface [1].

Once a facility has documentation of a pathogen in the environment, corrective action steps must be followed, documented, and verified. The inspector will review the documentation and review the reasons for and results of corrective action to remove the pathogen from the environment and investigate the root cause. The work that goes into corrective action is the vital part of an effective EMP. The first corrective action step is cleaning and sanitizing of the positive site. Sanitation probably has already been done at least once, if not more than once since sampling. If warranted, management may ask that equipment be further disassembled, more rigorously cleaned, and resanitized. Sanitation documentation should be reviewed along with SSOPs, with any deviations or unusual circumstances noted. Refresher SSOP training can be delivered as needed.

Investigational swabbing

Routine swabbing is what has been discussed to this point for the implementation of an EMP. Once a positive pathogen result is reported from an NFCS or a spike in count is seen from an indicator test, swabbing becomes investigational. As a consultant, my job is to assist companies with investigational swabbing and root cause determination as part of a recall investigation or risk assessment of the facility. In the event of a recall investigation, the line or facility will be shut down from production to facilitate swabbing of equipment, lines, and rooms. Hundreds of swab samples will be sent to a laboratory for testing. FDA may not allow production to resume until the root cause is determined and swabs test negative for the pathogen. The cost of the investigation and loss of production may reach into the millions of dollars. There are two tools I use in investigational swabbing after the report of a positive pathogen finding: vectoring and break down of equipment.

Vectoring

Imagine a pathogen is found at a site; let's use the example of a zone 2 site on a filling machine at packaging. From that positive site, investigational swabbing considers the potential source of the pathogen in three dimensions. How did the pathogen travel to the surface of the filling machine? Look above the site for a source of the pathogen that could have dropped or dripped down. Look horizontally around the site for potential sources. Look below the site. A drain in proximity to the site could send a contaminated aerosol up to the filling machine. Investigational swabbing collects 5–10 swab samples from the area around the positive site in the pursuit of the path to the positive site. This three-dimensional swabbing is called vectoring. Some people in the food industry refer to the "starburst" pattern of vectoring around a positive site. In-house investigators should take site samples logically to trace the source of contamination. Investigational swabbing is very costly, as each positive site leads to multiple vector swabs. When in turn a vector swab is reported as positive, additional vectoring of the second positive site means another round of swabs sent for testing. The process repeats until the root cause is understood, the area is cleaned and sanitized to remove the pathogen, and the pathogen is undetectable.

Break down of equipment

Sanitation crews follow SSOPs for disassembling equipment for cleaning and sanitizing. The interior surfaces of equipment are not part of routine swabbing in an EMP because interior surfaces are not exposed to potential environmental pathogens. However, during investigational swabbing the goal is to find the source of the pathogen, and the potential exists that the interior surfaces of the equipment may not be adequately cleaned and sanitized and may be harboring the pathogen. For this reason and during investigational swabbing, equipment will be shut down and disassembled for swabbing of interior surfaces. As a consultant, this is my favorite part. Working with maintenance for disassembly and operations for an understanding of the process flow, I stand at the ready to swab a gasket ring, filter, or parts of pumps as the maintenance employee aseptically opens the part. There is always the anticipation of what will be seen and learned from the once hidden surfaces of the equipment. If I am working on a recall investigation, the parts of the equipment will be disassembled down to the last bolt, screw, and plate, far beyond the disassembly by the sanitation crew. The goal is to discover where the pathogen has built a niche and harbors within. Our team is often told by maintenance that they can't remember if a given part has ever been taken apart or if a gasket has ever been changed, that the part will break, or that they won't be able to reassemble the part. Of course, those statements strengthen my resolve to get into the part!

As an edible safety professional, it is your responsibility to observe sanitation. Watch carefully how the equipment is disassembled for routine cleaning and sanitizing. Would it not be better to do your own investigation of niches within equipment parts before finding yourself in a recall investigation? A risk assessment is a focused research project to determine the effectiveness of sanitation. My personal opinion is that every facility should voluntarily conduct its own risk assessment annually in which equipment is disassembled beyond the routine SSOPs in the hunt for pathogen niches.

Summary

Cannabis-infused RTE edible manufacturers are encouraged to implement an EMP to verify the sanitation preventive control is controlling the hazard of environmental pathogens. Here is the reasoning in favor of environmental monitoring:

1. Environmental pathogens such as *Salmonella* and *L. monocytogenes* may be present in the processing and packaging rooms.
2. In the hazard analysis, cannabis-infused edibles manufacturers should identify the hazard of environmental pathogens at the post-kill process steps where the product is exposed and until the primary package is sealed.
3. The hazard of environmental pathogens is controlled with a sanitation preventive control.
4. The sanitation preventive control requires written SSOPs and records for monitoring, verification, correction, and corrective action.
5. Verification of the sanitation preventive control includes record review, use of dip strips, real-time swabbing, and testing surface samples for indicator microorganisms or pathogens.
6. Food facilities producing RTE products must have an EMP for the pathogens of concern as required by the FDA. With marijuana legalization, FDA can be expected to hold edible manufacturers to the same requirement as the food industry.

Environmental monitoring is common in the food industry for RTE products in the control of environmental pathogens. The development of an EMP takes time, resources, and management commitment. A cannabis-infused edibles manufacturer would be wise to implement an EMP for the dual purpose of protecting the immunocompromised consumer from pathogens and following the lead of the food industry before FDA regulation (Box 8.1).

Box 8.1 Real people in cannabis

Scott Hansen, Vice president of Quality and Regulatory Affairs, Agricor Laboratories/Botanacor Laboratories - Colorado.

Time in the cannabis industry: Agricor/Botanacor: August 2016–present.

Charlotte's Web: September 2014–August 2016.

What drives me in my job: After spending the first 20+ years of my career in the pharmaceutical industry, I was looking for a change and a challenge. In 2014, I interviewed with a start-up in Colorado's newly reintroduced hemp industry. As I walked through their office for the interview, I noticed photographs of children on the walls—children they wished to help with their hemp product—Charlotte's Web. I took the job and had the opportunity to speak with so many families and meet some of them in person, families with kids for whom traditional medicine had not helped—they were looking to us for hope. I was tasked with establishing the quality systems to ensure we made a safe and consistent product. Every quality decision we made we reflected upon the people we knew were relying upon us. Since then the hemp industry has exploded and there have been so many products, some which no doubt share the quality mission of that start-up but others who are out to make money without a concern for the quality of their products or the impact on the people consuming them. This led me to the next place that I could make a difference—the testing of these products. I joined Botanacor, one of the first cannabis laboratories focused on hemp in the country, with like-minded people from the pharmaceutical industry who wanted to ensure that any consumer could access clean, safe, and accurately labeled products. We worked with scientists and researchers to help understand processes and the baseline of the products in the industry, and we worked with regulators on statutes and rulemaking to ensure public safety. Every day we work with the manufacturers of products to help them understand why quality is so important to their consumers and their business.

References

[1] Control of *Listeria monocytogenes* in ready-to-eat foods: guidance for industry. Draft guidance. https://www.fda.gov/media/102633/download. [Accessed 11.02.19].

[2] BAM: food sampling/preparation of sample homogenate. https://www.fda.gov/food/laboratory-methods-food/bam-food-samplingpreparation-sample-homogenate. [Accessed 11.02.19].

[3] Taylor TM, Sofos JN, Bodnaruk P, Acuff G. Compendium of methods for the microbiological examination of foods. 5th ed. Washington, D.C.: American Public Health Association; 2015 [Chapter 2].

[4] The enterobacteriaceae and their significance to the food industry. https://ilsi.eu/wp-content/uploads/sites/3/2016/06/EP-Enterobacteriaceae.pdf. [Accessed 01.21.20].

[5] HQA indicator assay: a useful hygienic quality assay – HQA testing. http://new.kornackifoodsafety.com/?page_id=226. [Accessed 09.14.19].

[6] HTEB assay: a useful hygienic indicator assay – HTEB testing. http://new.kornackifoodsafety.com/?page_id=170. [Accessed 09.14.19].

[7] Food Safety and Preventive Controls Alliance. Preventive controls for human food. Peanut Butter Model Food Safety Plan. 2016. p. 13.

[8] Roggen M. Healthy and concentrated cannabis plants: how to use acronyms to optimize production. Journal of AOAC International 2019;102(2):421–6.

[9] Food Safety Modernization Act (FSMA). https://www.fda.gov/food/guidance-regulation-food-and-dietary-supplements/food-safety-modernization-act-fsma. [Accessed 11.02.19].

CHAPTER

Operation: manufacturing of edibles

9

Characteristics of cannabis-infused edibles

Cannabis-infused edibles are an option for the medical patient or recreational user who prefers not to smoke. The edible does not generate the smell of marijuana as is characteristic from smoking, and children and nonusers do not inhale the second-hand smoke. Edibles can be discretely consumed among nonusers. For the most part, edibles sold in the United States share the characteristics of being homogeneous for dosing and potency, conducive to packaging of individual serving sizes and being shelf-stable.

Homogeneous

Homogenization of the cannabis concentrate throughout the edible batch is critical for every serving to contain the same amount of cannabidiol (CBD) or Δ^9-tetrahydrocannabinol (THC). The consumer will read the amount of CBD and THC per serving on the label and calculate the number of edibles, e.g., number of gummies, or portion of an edible, e.g., part of a chocolate bar, to consume for the desired dose, i.e., the total quantity of CBD or THC to consume for the desired medicinal or recreational effect. Think of the process of making cannabis-infused edibles. The cannabis concentrate of known potency for CBD and THC is added to the batch. A cookie batter is stirred to smoothness. The syrup for gummies or lollipops is uniformly mixed. Melted chocolate is slowly mixed to blend both the concentrate and other flavorings. A batch of a beverage is blended before cans or bottles are filled. Edibles may be coated snacks in which the concentrate is homogenized into the coating. Most of our currently available edibles start by creating a homogenized batch that will be portioned with measured dollops or filled into forms or packaging of a given size. The portioning step is critical for achieving a uniform serving size for each edible.

There are exceptions to homogeneous edibles. Uniform, consistent dosing is more difficult to achieve in edibles with particulates. Chunky cookies and bars start with batters that are not smooth and homogeneous and make consistent dosing difficult. One way to achieve consistent dosing with chunky chocolate bars is to place the nuts or particulates in the chocolate mold first and then add the same volume of melted chocolate with the homogenized concentrate into each mold around the

Food Safety Lessons for Cannabis-Infused Edibles. https://doi.org/10.1016/B978-0-12-819512-3.00009-2

particulates. Another means to create chunky bakery items or chocolate bars is to start with one layer of a smooth mixture with the cannabis concentrate for the consistent dosing of each serving and place a second layer on top without the cannabis concentrate and with the chunky particulates. Cannabis chefs create chocolate-coated edibles in which the cannabis concentrate has been homogenized into either the center of the confection or into the chocolate coating for a consistent dose of CBD or THC.

Portion-evident or individually packaged

Chocolate bars are molded with portion-evident squares that can be easily broken apart from the bar, with one or two squares being the serving size. Chocolate bars and bags of gummies are examples of an edible in which multiple servings are sold, and it is left up to the consumer to calculate the portion to consume for the desired dose. It is a good test of willpower in the novice user to not consume the entire cannabis-infused chocolate bar. Experienced cannabis users can consume more than a novice user and enjoy the option of purchasing an edible with multiple servings and less packaging.

Individually packaged edibles are heavy on packaging and deliver a single serving with a small dose. Examples include small pieces of chocolate, one cookie, one gummy, or a lollipop. These individually packaged edibles are appropriate for the novice user who wants to easily control their portion and dose.

Shelf-stable

Shelf-stable products are held at room temperature and do not require refrigeration or freezing. The water activity of shelf-stable products is typically well below 0.6 and thus does not allow the spoilage of the product due to yeast or mold growth. The Cannabis Act in Canada requires all cannabis-infused edibles to be shelf-stable and allows dried meat products such as jerky. Most edibles in the United States are shelf-stable for the reason of sample testing programs, which require the edible to be held until laboratory results are released. Until the cannabis laboratory releases test results, companies must hold product from sale, but not all edibles are shelf-stable. A company is selling cannabis-infused ice cream in Massachusetts, a state with legal recreational use. The frozen edible is easily held in frozen storage until test results are released from the cannabis laboratory like edibles that are shelf stable are held at room temperature. Refrigerators and freezers are not commonly seen in a dispensary in contrast to being commonplace in a supermarket because most edibles are stored at room temperature.

Foods made into cannabis-infused edibles

Many states use the term marijuana-infused products (MIPs) instead of cannabis-infused edibles. Look for the synonyms in the state code. The following cannabis-infused edibles share the characteristics of being homogeneous, portion-evident, and shelf-stable. Cannabis-infused edibles go into product lines defined by the manufacturing and packaging equipment.

- **Baked goods**
  - Brownies
  - Bars
  - Cookies
- **Soft confections**
  - Gummies
  - Honey and honey sticks
  - Chews
  - Caramels
  - Marshmallows
- **Hard confections**
  - Lozenges
  - Lollipops
  - Mints
- **Chocolate**
  - Dark chocolate
  - Milk chocolate
  - Pieces
  - Bars
- **Snack mix**
- **Beverages**
- **Dry products**
  - Tea
  - Cocoa
  - Creamer

The Food Code and Food Safety Modernization Act rules are all under the jurisdiction of the FDA at the federal level. Edibles have the characteristics of a food. What else is under FDA enforcement? Pharmaceuticals and dietary supplements. Cannabis-infused edibles are not pharmaceuticals because they have not met rigorous drug testing. Cannabis-infused edibles could be classified as dietary supplements rather than as food by the FDA because of the dosing and labeling of CBD and THC.

Product development

Many commercially available foods started in the home kitchen. My father-in-law grew up drinking his mother's home-canned tomato juice. With her gone, V8 vegetable juice, always in his refrigerator, is the substitute for his mother's tomato juice. Many processed foods such as V8 can be traced back to their origin as a recipe from a home cook or chef. The story of the Nestle Toll House cookie recipe is delicious [1]. Lodging was provided at the Toll House Inn on the road from Boston to Cape Cod. In 1930, Ruth Wakefield, the innkeeper and cook for the travelers, chopped chocolate, added the chocolate bits to cookie batter, and voilà, chocolate chip cookies were invented!

We are still in a cupcake and cake decorating craze. There are cupcake caterers and stores dedicated to cupcakes; we see cupcakes at weddings instead of the traditional wedding cake. Interest in cake decorating led to popular TV shows like "Cake Boss" and "Ace of Cakes." I predict someday soon there will be a popular TV show on making cannabis-infused edibles. Wouldn't it be fun to see the behind-the-scenes action of an edible operation? Personally, I would watch a show on extraction or cannabis lab testing too, but I am a geek.

There are unique food products that started in someone's kitchen, maybe moved to a community commercial kitchen, and then grew into their own manufacturing space. One story like that is CLIF bars [2]. The CLIF bar was inspired in a 175-mile bike ride by Gary Erickson in 1990. The only type of bar available at that time was hard and dry. The CLIF bar was invented in the kitchen of Gary's mother.

Food science

The food industry employs product development scientists within product development departments creating a new product or changing a recipe of an existing processed food. Product development scientists are often educated as food scientists, like I am. Product development scientists in the food industry may be food scientists who learn culinary arts or skilled chefs who learn food science. Professionals who have worked as chefs make excellent food scientists in product development. The professional association, Research Chefs Association [3], is for chefs in the food industry, and I encourage cannabis chefs to join this professional association.

The working space of a product development scientist in the food industry is part chemistry laboratory and part kitchen. I have seen product development laboratories with the best, current models of in-home appliances that are beautiful working spaces. Typically, this laboratory or adjacent rooms function for sensory analysis, a significant food industry sector that employs food scientists. Sensory science uses human subjects, test equipment, and statistical analysis in the evaluation of new flavors and new products. Beyond the product development laboratory and sensory laboratory, scale-up of new products move to the pilot plant with small-scale processing equipment and where food and chemical engineers create small batches of the new product. In this way, potential problems in scale-up can be discovered before resources are spent on full-scale production of a product.

The transfer of skills from culinary to the food industry is valuable. The same is true in the cannabis industry. Does the cannabis industry know that food science exists? Probably not! Food science is not well known. Members of my own family think I work in nutrition or cooking. It must be my wonderful cooking skills that confuse them! Every cannabis-infused edible manufacturer would be wise to hire its first food scientist and learn the value of the technical resource of a food scientist.

Cannabis science

Within cannabis science (I like the sound of that term), processing and packaging rooms become kitchens and scientists are chefs. As companies embark on construction for their edible production, a build-out in the cannabis-infused edibles industry is called a "kitchen build-out." The person in charge of product development is referred to as a cannabis chef and often is, in fact, someone skilled as a professional chef. The cannabis chef will create recipes for the edibles and test batches of the recipe over and over in preparation for launching the edible. I was very confused when I first heard people in the cannabis industry refer to manufacturing in the "kitchen." I finally figured out that the term for the manufacturing area in a cannabis-infused edible facility is the "kitchen." When I see a video of edible production or pictures on websites, the manufacturing environment looks like a commercial kitchen, so I understand the use of the term. However, for the success of cannabis companies making large volumes of edibles, large-scale production must be constructed and should emulate the space used by large bakers, confectioners, and chocolate makers in the food industry. The processing and packaging rooms in the food industry are not called kitchens. Many cannabis kitchens are the size of product development kitchens. The cannabis industry needs to keep the cannabis kitchen as a product development laboratory as in the food industry, build pilot plants for scale-up, and then build full production facilities on par with the food industry. Aspire to be the next Hershey, Jelly Belly or Nabisco of the cannabis industry.

Homogenization of cannabidiol and Δ^9-tetrahydrocannabinol, again

When it comes to the manufacture of an edible, there is very little difference from the manufacture of the traditional, noncannabis food product. The ingredient bringing in CBD or THC becomes an ingredient in the recipe and factored into the batch. As an oil or butter, there will be functional properties in the edible of the oil or butter and considerations for dispersion. However, one difference from traditional food processing is that before batching of ingredients and mixing, the potency of the ingredient with CBD and THC must be known. If the cannabis-infused edible manufacturer is purchasing the ingredient or if the manufacturer is making the ingredient themselves, the ingredient must be tested for the CBD or THC concentration per state code. When batching, this concentration can be factored in to get the right dosing into the batch and potency per serving. One of the key steps in cannabis-infused edibles manufacturing is homogenization of the CBD or THC ingredient into the batch. After the edible is made, the potency of the product must also be tested for the CBD or THC concentration per state code. The manufacturer must follow state regulations for sampling programs and transport to the cannabis laboratory. Once the manufacturer receives the test results, they can move on to labeling.

The edible will be labeled with the potency of CBD and THC per serving size in a similar way in which ingredients are labeled in the Supplement Facts of a dietary supplement. Dietary supplement manufacturers are required to follow FDA current Good Manufacturing Practices in 21 CFR 111 [4]. How will the FDA classify and regulate cannabis-infused edibles? My hope is that the FDA will use current regulations for pharmaceuticals, dietary supplements, food, pet food, and tobacco products and take the best parts for compliance and enforcement. For edibles, it would mean following GMPs from 21 CFR 111 and 21 CFR 117; writing and implementing an edible safety plan addressing supply chain, process, sanitation, and allergen preventive controls; and labeling with the requirements of a dietary supplement with Supplement Facts. Edibles currently **cannot** be labeled with Nutrition Facts or Supplement Facts because those are part of the federal law for food and dietary supplements, respectively, and cannabis is illegal at the federal level. With no federal oversight, edibles are not pharmaceuticals, food, or dietary supplements. Edibles manufacturers must follow the labeling requirements written into each state code.

Build food safety into product development

Edible safety plans are food safety plans and typically written based on the equipment and common hazards for a product line. If chocolate making uses all the same equipment for all the flavors, sizes, and shapes of the molds, then all the chocolate bars can go on the same food safety plan. If there is a unique allergen, such as a nut, in one bar but not in the others, a single food safety plan may incorporate the unique allergen, or the allergen may be addressed as an optional part of the food safety plan. FDA allows flexibility in the control of hazards. Baked goods can be grouped together as can soft confections such as gummies. By grouping a product line into one food safety plan, the common hazards are addressed independent of individual products. It is important in the development of a recipe by a cannabis chef that edible safety is built in from the start. A new product in an established line can probably be added to an existing food safety plan for the edible. Starting a completely new line of edibles probably is the start of a new food safety plan. Time and temperature of the kill step, if there is a kill step, and sourcing of safe ingredients are critical for the safety of the edible from the start of product development.

Validation of process preventive controls

If the edible is a gummy or hard candy, the cooking time and temperature create a kill step to control pathogens. If the edible is a baked good and after portioning, the time and temperature of baking will be monitored for the kill step. I have not heard a great deal of discussion of validation of kill steps in the cannabis industry as is required by the FDA for food manufacturers. Edible manufacturers are living in one of two worlds, depending on their state code. If the state code requires a Hazard Analysis and Critical Control Point (HACCP) plan, the state may be enforcing food safety on a cannabis-infused edibles manufacturer in the same way they enforce food safety on a restaurant. The cannabis company would follow the state or federal FDA Food Code for time and temperature values for cooking. The Food Code is written for food service, and cannabis-infused edible manufacturers are not restaurants, cafeterias, or grocery stores.

If not living in the world of food service, cannabis-infused edible manufacturers may be defined as food facilities, i.e., receiving facilities as defined by the FDA Preventive Controls for Human Food (PCHF) rule [5]. Cannabis-infused edible manufacturers must control hazards beyond HACCP with supply chain preventive controls, sanitation preventive controls, or allergen preventive controls. When states adopt regulation that aligns with the FDA PCHF rule, 21 CFR 117, validation of process preventive controls is required. The good news is that the federal FDA Food Code can be cited in a validation study in support of cooking times and temperatures, as appropriate.

Supply chain preventive controls

There is no kill step in the making of cannabis-infused chocolate by melting bars or discs and adding a cannabis concentrate and flavors. With no kill step, there is no process preventive control. Control of biological hazards is instead at the supplier of the chocolate and other ingredients, including the cannabis concentrate. Purchasing of chocolate and other ingredients where the supplier controls the hazard requires a supply chain preventive control. Sourcing of safe ingredients from suppliers and implementation at the edible facility of effective sanitation, excellent personal hygiene, and use of personal protective equipment will be keys to a safe edible.

Over time and like the food industry, customers will drive product testing and environmental monitoring. Certification and auditing will be strong in the cannabis industry as it is in the food industry and drive the safety of cannabis products. Testing of product is dictated by the state code in the cannabis industry. In some cases, the state code is unreasonable in my opinion and in some ways lacking. The cannabis industry can learn from the food industry about the weakness of product testing and how to build environmental monitoring into the process of manufacture of edibles. I do like that the state code requires testing of the concentrate as an ingredient. In this way, the state code has required a supply chain preventive control on the concentrate as an ingredient in the edible, and that makes sense.

Written standard operating procedures

There are so many details to implement in manufacturing. Written standard operating procedures (SOPs) document the procedures used in a facility with the supplies, materials, and equipment on-site. SOPs will be important for any certification scheme that is pursued and used in the training of personnel. By following a written SOP, every employee has the same directions for a procedure. The following is a short list of SOPs for consideration at an edible facility.

- Receiving
- Dry goods storage
- Secure storage of the concentrate
- Shoes and boots
- Clothes and uniforms
- Handwashing
- Personal protective equipment
- Batching
- Cleaning of utensils, tubs, and trays
- Cleaning of equipment and rooms
- Master sanitation schedule
- Pest management program
- Allergen management program
- Secure storage of product on hold
- Release of product for packaging and labeling
- Labeling of packaged product
- Sampling plan
- Transportation of samples to the cannabis laboratory
- Transportation of labeled product to the dispensary

Every SOP for consideration here focuses on product safety. Each facility will generate additional policies and procedures for worker safety, payroll, hiring practices, and others outside the focus of product safety. For control of documentation, the policies and procedures for product safety should be stored separately from those not of product safety. While not endorsing, I like what I have seen of the software from CannaChek [6] to assist the cannabis industry in writing SOPs as an example of SOP management software.

References

[1] Toll house cookies: a long secret history (with recipes). https://the-toast.net/2014/12/05/toll-house-cookies-secret-history/. [Accessed 11.07.19].

[2] How we got started. https://www.clifbar.com/article/how-we-got-started. [Accessed 09.14.19].

[3] Research chefs association. https://www.culinology.org/. [Accessed 09.14.19].

[4] 21 CFR 111, Current good manufacturing practice in manufacturing, packaging, labeling, or holding operations for dietary supplements. https://www.accessdata.fda.gov/scripts/cdrh/cfdocs/cfcfr/cfrsearch.cfm?cfrpart=111. [Accessed 09.14.19].

[5] FSMA Final rule for preventive controls for human food. https://www.fda.gov/food/food-safety-modernization-act-fsma/fsma-final-rule-preventive-controls-human-food. [Accessed 09.14.19].

[6] Quality management for cannabis production. https://ironapple.net/iron-apple-qms/cannabis-qms/?CANNACHEK. [Accessed 12.06.19].

CHAPTER

Operation: packaging and labeling of edibles

10

The role of packaging

I really geek out when I see fillers and packaging equipment. I should have been an engineer, but nobody guided me in that direction, and I learned of that direction too late. Packaging equipment and I have a relationship. I have disassembled many fillers and package sealers in the hunt for pathogens, and I have great respect for the design and use of filling and packaging equipment, along with the engineering feat of product movement from one processing step to the next. It is fascinating. The cannabis flower filler in the video in Ref. [1] is a thing of beauty and probably brand new because it is extremely shiny and unscratched. Following 1 month of use, I could disassemble parts of the filler and show you places that have never been cleaned; that's what I do. I am happy to see such pieces of equipment in common use in the food industry are crossing over to the cannabis industry. With example after example, the same equipment is used for chocolates, gummies, lollipops, cookies, and beverages. Crossover exists from the dietary supplement industry as well. Equipment from the dietary supplement industry for filling capsules and compressing tablets is used in the cannabis industry for making edibles in the form of dietary supplements. It just makes sense that what works in one industry should work in another industry.

Colorado field trip

I purchased a small box of cannabis-infused chocolate in Colorado, like one of those mini Valentine's hearts that have three pieces of chocolate. I don't have the receipt but the box of three chocolate pieces cost about $15, more than the cost of exquisite chocolate. I could not see the chocolate at the point of sale, which was disappointing because I love and enjoy chocolate; the box was labeled 15 mg of Δ^9-tetrahydrocannabinol (THC). That night I went to have some chocolate and I could not open the box. Period. To this day I do not know how that box was meant to be opened. So, I tore the box open. Talk about tearing into a cannabis product! The box was beautiful, and inside were three unique pieces of chocolate. Imagine three precious pieces of fine chocolate each in a different geometric shape and each resting in their own spot in a single formed cup holder that filled the area inside the box. Given the cost, I paused to take in the beauty of the chocolate. I ate one piece of chocolate a little too quickly

Food Safety Lessons for Cannabis-Infused Edibles. https://doi.org/10.1016/B978-0-12-819512-3.00010-9

because given the cost I should have savored the flavor. I was on a research mission to experience the effects on me. It takes about an hour to feel the effects of an edible, so being a scientist, I noted the time and hunkered down to watch TV in my hotel room. Then I panicked. I did not know how much THC a novice should consume. I called my husband; I googled. I learned that 15 mg of THC is too much to start with, and the recommendation was 5 mg. Just great. Note: there are no reported deaths from consuming too much THC. Then I went back to the torn outer packaging of the chocolates. Reading cannabis-infused product labeling is difficult, but that may be because I am new to the industry, so buyer beware. Yes, the box was labeled 15 mg. Somehow from the label information I deduced that each piece was 5 mg. Whew. Catastrophe averted. It took about an hour and a half for me to feel anything. I became a little light-headed or woozy, which lasted 5 minutes. The other two pieces of chocolate were left for housekeeping. Warning: do not, I repeat do not, consume more cannabis-infused edibles when you don't feel the effect from the first ingestion. Although you won't die from an overdose, you may have a bad experience or do something crazy and end up harming yourself. As I relay this story to my cannabis colleagues, I am told that it is common not to feel the effect of cannabis with many of the first trials or experiments.

Important points about packaging of cannabis-infused edibles

Packaging must be designed to be [2]

- functional and opaque,
- portioned or individually wrapped,
- tamper-evident,
- child-resistant,
- unattractive to minors,
- unlike certain commercially available products,
- a protection of the contents from contamination.

Functional and opaque

Control of water activity

Packaging engineers have great job opportunities in the cannabis industry for package design to creatively meet the packaging requirements found in each state code. Boxes, pouches, and foil wrap are used to cover and protect edibles. When the edible is a low-water-activity product as most edibles are, packaging of cannabis-infused edibles must function in the protection of the edible from humidity. Addition of moisture into edibles such as gummies, chocolates, and bakery items will initiate organoleptic changes and microbial spoilage. If a pathogen is in the edible, the water activity may no longer function to prevent growth. I concede that some edibles have an extremely low water activity in that the exposure to humidity will not affect the quality or safety of the product; hard confections such as lollipops are an example.

Opacity

By state code, packaging must be opaque. A functional advantage of opaque packaging is the blocking of light. Colors remain vibrant, and light-mediated chemical spoilage of product is reduced. If an edible is packaged such that you can see the product, the packaging is illegal. Looking at the edible before buying is not allowed, and that used to drive me crazy. Part of the organoleptic experience is using the sense of sight to see what I am purchasing, and packaging of edibles does not allow the consumer to inspect the edible before buying. Buying a cannabis product is like buying a mystery shopping bag at the grocery store. Here again, it was my novice approach to buying edibles that led me down the wrong path. One day I was thinking about the requirement for packaging of edibles to be opaque, and I started to think about the packaging of food. I could not think of one chocolate bar that is sold in transparent packaging. Can you? That is not the story with gummies, as often gummy bears and gummy worms are seen through the bag. Canned beverages are not seen until poured. Most nationally and commercially available bakery items are unseen until unwrapped. I have settled into the understanding that familiar processed and packaged food is typically unseen at the time of purchase, so I am now comfortable with the opacity requirement for packaging of edibles.

Portioned or individually wrapped

Cannabis-infused edibles and their packaging must be designed to be either portioned or individually wrapped to provide the labeled dose per serving size. Some cannabis-infused edibles such as chocolate bars or bakery items are meant to be multiserving edibles, with each serving delivering the labeled dose. However, what if mom or dad opens a multiserving chocolate bar and consumes one serving? Now the remainder of the bar with up to nine servings is left to store safely away from children. Ideally, the packaging would be resealed to demonstrate both tamper evidence and child resistance. A packaging engineer may have a solution, but for me, it is difficult to imagine a chocolate bar wrapper that is child-resistant yet alone can be resealed. Instead, the rewrapped chocolate bar could be safely stored at home in a child-resistant canister or maybe a safe! Education campaigns on the safe storage of edibles compliment the use of child-resistant packaging [2]. Many edible manufacturers have addressed the concern of storing opened edibles by using tamper-evident packaging for individual serving sizes of edibles and not selling multiserving edibles.

Tamper-evident

Cannabis products are heavily packaged, in general, and meant to be tamper-evident. As I experienced with chocolate in Colorado, I literally had to tear the box open to access the chocolates, which is tamper evidence to the extreme. In the process of protecting the product with packaging and making the packaging tamper-evident, the packaging engineers may have gone a bit too far. There is a lot of packaging material used in the cannabis industry, including plastic inserts, foil wraps, paperboard containers, stickers to seal the container, and plastic film like that used on the outside of cigarette packs. By the time the package is opened to the cannabis product, the product sometimes appears small and insignificant. The benefit of all this packaging is tamper evidence and child resistance.

Child-resistant

Opacity is a feature of child-resistant packaging so that the allure of the edible is not experienced by the child. Child-resistant packaging involves the design of opacity, the opening mechanism, use of warning labels, and testing of packaging for child resistance [3]. One of the reasons that edibles are often sold individually wrapped in single serving sizes is the requirement for child-resistant packaging. For the consumer, the cost to purchase single servings of an edible is greater than that for a multiserving edible because of the greater amount of packaging. This is a disadvantage to many consumers with greater tolerance who consume more than one serving for the desired effect and prefer a multiserving edible. The inconvenience experienced by adults pales in comparison to meeting the requirement for child-resistant packaging.

A 2016 survey by the Council on Responsible Cannabis Regulation in collaboration with the National Cannabis Industry Association of 178 respondents found the majority supported child-resistant packaging not only for edibles but also for cannabis flower, concentrates, and noningestible infused products [2]. Flower or concentrates that have not been decarboxylated will not produce a psychoactive effect if consumed, but the collaborators decided to be on the safe side by recommending child-resistant packaging for all cannabis products [2].

Unattractive to minors

Packaging of cannabis-infused edibles cannot be attractive to children, including all minors. "Minor" means a person under the age of 21 years for adult-use cannabis or under the age of 18 years for medical cannabis [2]. The cannabis industry, like the manufacturers of alcoholic beverages and cigarettes, responsibly market and sell to adults. There will be no toy in the box of edibles like there is with Cracker Jack. There will be no children's movie sponsorship like there is for the toys in and packaging of the Happy Meal at McDonald's. In contrast to packaging that markets to children, edibles packaging designed for the adult market can be elegant. Look at the packaging for Darwin Brands of Arizona [4]; the design is not only for adults but also masculine. I love the art and design of the packaging for Défoncé in California [5]. You will be hungry for chocolate after perusing the website of Mindy's Edibles [6] from James Beard Award–winning Chef Mindy Segal in Illinois. The likes of Darwin, Défoncé, and Mindy's make you feel all grown up when you are purchasing edibles for yourself. Like a fine wine, an edible is meant to be savored and enjoyed by adults, not children.

Edibles and their packaging unlike certain commercially available products

Most state code specifies that edibles are not allowed to be brightly colored or of shapes attractive to children. Specific shapes may or may not be listed in the state code; for example, there should be no cannabis-infused gummy bears or gummy worms because those are common candy shapes for children. When a cannabis-infused edible is being developed, it is critical for the company to know and apply the state code for design of color and shape of the edible and transfer the design into production with appropriate molds and packaging equipment. The design is also critical for the packaging material colors and shape to prevent the accidental ingestion of a cannabis-infused edible by a minor. Packaging of cannabis-infused edibles cannot mimic a traditional noncannabis product. If I say TicTac, you know the packaging of the product. If I say REESE's Peanut Butter Cup, you know the design of the packaging and shape of the product. Creating a product name, product, or packaging that mimics a commercially available product is irresponsible and will get a cannabis company in trouble.

A protection of the contents from contamination

The final feature of packaging for cannabis-infused edibles makes my heart sing as a food microbiologist. Food safety is always a priority, and the design of packaging must include the prevention of cross-contamination with a pathogen through package integrity. Likewise, the edible contact packaging material must not transfer harmful chemicals causing the edible to be adulterated. As always, find and follow the state code in designing and meeting requirements for packaging. The state of California provides a nice checklist of packaging requirements [7] and labeling requirements [8] for cannabis products.

The role of labeling

Labeling is applied to packaging for the communication of information to the consumer about the cannabis company and edible. Information that is recurring such as product name, company name, and edible weight or volume may be preprinted on the outer packaging material. The cannabis industry employs the use of stickers to detail information such as cannabis strain and potency that may change from batch to batch. The cannabis industry operates on a hold-and-test program where the cannabis product is made and held until test results are released from a cannabis laboratory. The state code provides tolerance around the labeled potency. For example, if a product is labeled as 100 mg of THC and the state has a 10% tolerance, the potency can range from 90 to 110 mg of THC. Once the potency is reported from a cannabis laboratory, the cannabis product can be brought from storage and labeled. Unlike a typical Nutrition Facts label for a food or a Supplement Facts label for a dietary supplement that is directly printed on the packaging, a sticker with information specific to the batch of the cannabis product will often be applied to the package of the cannabis product. This sticker is not attractive; the sticker is informative and provides compliance with the state code.

Nutrition Facts and Supplement Facts

The FDA Nutrition Labeling and Education Act details the requirement for Nutrition Facts labeling of foods [9]. In 21 CFR 101 Subpart C Section 101.36, FDA details the labeling requirements for dietary supplements and format of the Supplement Facts label [10]. The Supplement Facts label does not include Reference Daily Intakes or Daily Reference Values as found in the Nutrition Facts label [11]. I am concerned about cannabis companies who create a Nutrition Facts or Supplement Facts label for their cannabis-infused edible thinking they are doing the right thing by informing the consumer about the nutrition of the edible. However, cannabis-infused edibles are neither a food nor a dietary supplement by definition at the federal level. FDA heavily regulates the Nutrition Facts and Supplement Facts label, and FDA has been very clear on its stance that hemp-derived cannabidiol (CBD) edibles are illegal [12] as either a food or a dietary supplement. My concern is that a cannabis company applying a Nutrition Facts or Supplement Facts label on its packaging may garner the unwelcomed attention of FDA and make trouble for itself. A company may label the nutritional content of the edible on the label without the formatting of the Nutrition Facts label.

A Cannabis Facts label offers an alternative to a Nutrition Facts or a Supplement Facts label and may be seen on some cannabis products [2]. The Cannabis Facts label communicates the cannabis active ingredients such as THC and CBD, communicates their content by percentage, and could also detail the strain, testing date, cannabis laboratory, or other relevant information [13].

Allergen declaration

Despite my reasonable argument against the use of the Nutrition Facts label, the state of Oregon requires the labeling of edibles with a Nutrition Facts label and offers three formats of choice that must include nutrition information, serving size, number of servings per container, list of ingredients, and allergen information [14]. The primary purpose of the Nutrition Facts label is to inform the consumer of the nutritional content of the edible per serving size, putting the edible in the same category as a food with a Nutrition Facts label.

In addition to the Nutrition Facts label, an allergen declaration is required by Oregon to inform consumers who are allergic to an ingredient [14]. I absolutely support the use of an allergen declaration on edible labels for chocolates, bakery items, and other edibles that are known to contain allergens. The FDA requires allergen declaration for the big 8 allergens of wheat, soy, milk, eggs, peanut, tree nut, fish and shellfish. Countries outside the U.S. may list more allergens to declare. The allergen declaration is critical for informing the allergic consumer about the presence of ingredients known to be allergenic. Precautionary labeling is different from the allergen declaration but is also heavily used in the food industry to warn consumers that a product is made in a facility in the presence of undeclared allergenic ingredients. For example, I read the precautionary labeling on a cereal box, "made on a line that also has peanuts," when there was no peanut ingredient in the cereal. Seeing what I have seen of equipment in the food industry of difficult to clean equipment gives respect to the company that chose the precautionary labeling. The allergic consumer is informed of the potential for cross-contact of the peanut allergen, and the precautionary labeling allows the consumer to make an informed decision about consuming the product.

Health claims are made for products that have been tested robustly by large, expensive research projects. There are no recognized health claims for CBD or THC, and FDA is sending warning letters to cannabis companies that make health claims with regard to specific conditions such as cancer, arthritis, or anxiety [15]. Epidiolex is the only cannabis pharmaceutical that has been approved by the FDA [16], which contains CBD. All CBD edibles and dietary supplements with CBD are illegal in FDA's eyes [12].

The role of security

Food defense resembles security in the cannabis industry

My journey in food defense started in 2002. After September 11, 2001, my colleague, Matt Botos, as director of the Illinois Center for Food Safety and Technology, designed and implemented food defense training for the food industry. As outreach manager at the National Center for Food Safety and Technology, I supported the training. Matt was way ahead of where the food industry is now with food defense. In 2016, I became a lead instructor for the Preventive Controls for Human Food (PCHF) rule and followed the publishing of the final Food Safety Modernization Act rule on Intentional Adulteration or food defense to mention food defense in the delivery of the PCHF workshop. In 2017 and 2018, I received training in food defense at the Food Protection and Defense Institute in Minneapolis, a Center of Excellence in the federal Department of Homeland Security, and in 2018, I started delivering food defense training for NSF International as a webinar and on-site with clients.

Since 2017 I have been dabbling in food defense, as someone with food safety expertise. As a food scientist, you may have been drawn into food defense training too or even be on your facility's food defense team. At first, I thought food defense would be a natural extension of my skill set as a food microbiologist. Think anthrax, Ebola, and germ warfare. What I have learned is that food defense is a whole different animal, and I am thankful to the people in the food industry who keep our ingredients and products safe from attack.

The basics of food defense in the food industry include control of access to the facility by vehicles and people, identification of employees, cameras inside and out, receiving procedures, and control of access to ingredient storage, equipment, and processing. In the food industry, there is confidence that food defense is accomplished, once the product is in a sealed package and filled into a box and the box is placed in a case with the pallet of cases shrink-wrapped. There are four layers of packaging for protection.

I worked with a client to perform a gap assessment in allergen management. The client has a twin facility where one side manufactures the product with nuts and the other side is nut-free. The client has built two physically separated sides of the facility that share a common wall. In the review of employee foot traffic, I was pleased to learn that the client's key card access system only allows scheduled employees to enter the nut-free side for work. It is not possible for an employee to finish working on the nut side and move to the nut-free side. This is a system that could be applied throughout the food industry to control employee and visitor foot traffic and access to restricted areas. Despite all the training I received in food defense, the best training was to see defense applied at cannabis facilities. Compared with the cannabis industry, food defense in the food industry is child's play.

Security in the cannabis industry

Security in the cannabis industry goes beyond packaging and provides lessons for the food industry. I can only speak to my personal experience in visiting cannabis facilities, and this is what I know. Finding the facility is tricky because signage is limited, and the facility is not advertising its location, even if it is a dispensary. Growing operations are usually unmarked. When I visited facilities, I was provided with an address, and most facilities did not have a business sign or address number on the building. Approaching the building, there will be only one option for a visitor to enter. Some facilities use the visitor entrance for employees or some employees; otherwise, there will be only one employee entrance. All other doors and bays will be locked from the outside, but will be functional to meet fire code. Aside from the visitor entrance, the perimeter may be fenced with a locked gate. There are security guards; the guards may or may not be armed. The guard may be outside the building at the visitor entrance, in the vestibule inside, or at the Shipping and Receiving bay. Imagine an armed guard stationed at a food facility Receiving bay for food ingredients and at Shipping for the food product. The cannabis industry is prepared, and the ingredients and products are protected.

Transporting cannabis to a facility or cannabis product from a facility takes two drivers. It is the buddy system of food defense applied to transportation. There is a significantly less likelihood of losing cannabis or cannabis products if there are two drivers. Imagine if transport of ingredients or food product required the payroll for two drivers instead of one.

There are cameras everywhere outside and inside. To implement food defense, food facilities can use the cameras that already are in place or can install more. In the food industry, I expect a facility to have dead spots outside the view of cameras. In the cannabis industry, I would expect cameras to view every area of the facility inside and out and there are also multiple cameras capturing different angles in some rooms. For example, I observed a cannabis packaging room with at least three cameras at different angles focused on the activities of handling and packaging product. The facility only conducted packaging and labeling with an on-site dispensary and had a total of 16 cameras (Box 10.1).

Box 10.1 Myth

People in the cannabis industry get free weed

No! Employees in the cannabis industry may get a discount at a dispensary affiliated with their company. Employees do not get free weed and it is also written into the law that employees may not be given cannabis products as incentives, bonuses, or wages. This is to discourage corruption and bribes for dishonest work.

Once inside a cannabis facility and standing at the reception desk, I was not allowed to sign-in until my escort verbally gave the ok. Even though I was escorted by management, I had to provide my laminated identification; I used my driver's license. I was told that a passport is acceptable too. I was provided with a laminated visitor tag, as many of us have in the food industry. At cannabis facilities, I was never alone except for bathroom breaks. They take escorting of visitors very seriously. As a consultant, I have been involved in multiple-day recall investigations or risk assessment swabbing projects. After the first few days of being escorted, I am often left alone in a processing area or unescorted as I walk through the facility. Never leave a visitor alone! I understand that the manager's tasks are piling up and there are pressing questions to answer. It is so easy for a manager to step away and leave a visitor for a short time. There is the safety factor; I could injure myself. Then there is the food defense factor; you don't know me. Personally, I do not have the mind-set of an attacker, but an attacker could pose as a visitor.

The other thing the cannabis industry takes very seriously is donning personal protective equipment (PPE). The all-disposable set of PPE included a hair net, surgical face mask, below-the-knee coat, booties, and gloves in the growing operation! In my mind the greenhouse employees were agricultural workers ready to enter a BSL-2 laboratory. If you are looking to invest in the cannabis industry, then invest in the disposable PPE industry too! I saw the use of extensive PPE at every step in the cannabis operations. As a visitor, I was all-in with all the PPE just to walk through operations.

Skipping to packaging of cannabis products, there is a lot of packaging material around an edible and great waste of packaging material by the consumer. Many cannabis products are single-use or single serving creating a high ratio of packaging to product. If you are a packaging engineer, you would love the cannabis industry! There are many details about packaging and labeling of cannabis products in the code of each state. If you are in regulatory affairs, you would also love the cannabis industry! If the food industry was regulated for packaging in the same way the cannabis industry is regulated, our food would be much more expensive. One thing is for sure, tampering with cannabis products is very difficult not only due to the cameras at the facilities but also secondarily due to extensive packaging (Box 10.2).

Box 10.2 Real people in cannabis

Robert Brodnick, PhD, MS, Titan Analytical, founder/CEO, headquartered in California. Also the founder/CSO of PhenoTerps Flavor House.

Time in the cannabis industry: In the current company, 2 years. Time in the legal cannabis industry, 5 years.

What drives me in my job: I am driven by the potential for cannabis to have a positive impact on Veteran health and wellness and its role in combating the opioid crisis. Before cannabis and/or hemp products are broadly considered as a treatment to reduce veteran suicides or to help those addicted to opioids find an alternative pain therapy, the industry's testing laboratories must take decisive action to incorporate lessons learned and best practices from clinical, forensic, pharmaceutical, and food testing arenas in order to give the public and federal regulators the confidence to help usher cannabis into the mainstream. Unbeknownst to me at the time, I began preparing for a career as a scientist in the cannabis industry while serving as a US Army Medical Services Corps officer. After gaining valuable experience managing and executing high-priority chemical/biological defense and combat casualty care clinical and preclinical investigations, my focus turned to operating and auditing high-throughput forensic counter-narcotics laboratories, where I frequently testified in Federal UCMJ courts as an Expert Witness on cannabis, its impact on human performance and behavior, and data defensibility in analytical testing laboratories. After reaching the rank of Major and completing multiple assignments, including a research tour of duty in Afghanistan in support of Operation Enduring Freedom, I transitioned out of uniform to take charge of the Quality and Consulting departments at Food Safety Net Services (FSNS). My time as the FSNS chief scientific officer was well spent, fortifying my experience in food microbiology testing and becoming a Food Safety Modernization Act (FSMA)-compliant Preventive Controls Qualified Individual. Most importantly, I developed a profound appreciation for best-in-class customer service and the culture of urgent supply chain risk empathy that is critical for success in analytical food testing, as it is becoming cannabis. We launched Titan Analytical with the mission of improving the safety and quality of cannabis, hemp, and food testing services available to the market by combining defensible analytical chemistry and microbiology, modern food safety principles, and thought leadership in quality-by-design testing laboratory operations. Better laboratories lead to safer cannabis and food for a better world.

References

[1] Canapa Solutions. Cannabis packaging machine for weigh filling into Containers. https://www.youtube.com/watch?v=4YQ_710BTIY. [Accessed 09.14.19].

[2] Council on Responsible Cannabis Regulation in collaboration with the National Cannabis Industry Association. Cannabis packaging and labeling: regulatory recommendations for states and nations. https://mass-cannabis-control.com/wp-content/uploads/2017/12/111717CannabisPackagingandLabelingRegulatoryRecommendationsforStates.pdf. [Accessed 09.14.19].

[3] 4 Key components of child resistant packaging. https://www.johnsbyrne.com/blog/4-key-components-of-child-resistant-packaging/. [Accessed 09.14.19].

[4] Darwin. https://darwinbrands.com/. [Accessed 09.14.19].

[5] Defonce. https://www.defonce.com/. [Accessed 09.14.19].

[6] Mindy's Edibles. https://www.mindysedibles.com/. [Accessed 09.14.19].

[7] Packaging requirements. Cannabis and cannabis products. https://www.cdph.ca.gov/Programs/CEH/DFDCS/MCSB/CDPH%20Document%20Library/PackagingChecklist.pdf. [Accessed 09.14.19].

[8] Labeling requirements manufactured cannabis products. https://www.cdph.ca.gov/Programs/CEH/DFDCS/MCSB/CDPH%20Document%20Library/LabelingChecklist-Products.pdf. [Accessed 09.14.19].

[9] Nutrition facts label. https://www.fda.gov/media/97153/download. [Accessed 11.13.19].

[10] 21 CFR PART 101 – FOOD LABELING Subpart C – Specific nutrition labeling requirements and guidelines Sec. 101.36 – Nutrition labeling of dietary supplements. https://www.accessdata.fda.gov/scripts/cdrh/cfdocs/cfcfr/CFRSearch.cfm?fr=101.36. [Accessed 11.13.19].

[11] Dietary supplement labeling guide: Chapter IV. Nutrition labeling. https://www.fda.gov/food/dietary-supplements-guidance-documents-regulatory-information/dietary-supplement-labeling-guide-chapter-iv-nutrition-labeling. [Accessed 11.13.19].

[12] FDA chief clarifies enforcement priorities for CBD products. https://www.marijuanamoment.net/fda-chief-clarifies-enforcement-priorities-for-cbd-products/. [Accessed 09.01.19].

[13] Cannabis facts JB label. https://www.johnsbyrne.com/blog/4-key-components-of-child-resistant-packaging/cannabis-facts-jb-picture/. [Accessed 09.14.19].

[14] Packaging and labeling pre-approval. https://www.oregon.gov/olcc/marijuana/Pages/PackagingLabelingPreApproval.aspx. [Accessed 09.14.19].

[15] Warning letters and test results for cannabidiol-related products. https://www.fda.gov/news-events/public-health-focus/warning-letters-and-test-results-cannabidiol-related-products. [Accessed 11.13.19].

[16] FDA approves first drug comprised of an active ingredient derived from marijuana to treat rare, severe forms of epilepsy. https://www.fda.gov/news-events/press-announcements/fda-approves-first-drug-comprised-active-ingredient-derived-marijuana-treat-rare-severe-forms. [Accessed 09.15.19].

CHAPTER

Training needed in the cannabis industry

11

Importance of training

Training in the cannabis industry is needed for a company's long-term survival. If you are in the cannabis industry to make a quick buck, you are in the wrong industry. This industry is trying to establish itself. Every day I see articles about how the cannabis industry is failing and not making money. The cannabis industry is not failing and is bringing in revenue by the boatloads. The question is what companies are making a profit? Companies that will survive this period in history have the long view; they have their goals for and eyes on the long game. They are building their team this year to grow the company so that next year they can make a profit.

There are three levels of training needed in the cannabis industry: (1) all employees, (2) supervisors, and (3) management, and resources are available for training at each level. Analogous to the training needed for every employee in the cannabis industry, the FDA requires food safety training for every employee working in the food industry to their role and defines a qualified individual as a person having adequate and appropriate education, training, or experience or a combination thereof for the duties of their role in a food facility. By applying the same definition to employees in cannabis, some candidates will bring a formal degree of education from a university or technical college; some candidates complete training through workshops, courses, or certificate programs; and some will bring on-the-job training and experience. There is not a single path, and each aspect of education, training, and experience for a cannabis employee is notable and valuable to an employer. Job candidates should be cautioned about adding black market experience on a resume in this professional field under regulation. It seems that putting black market experience on a resume would be like me putting on my resume that I have hosted parties and nobody got sick from the food I served. Not causing foodborne illness is a personal goal in my life but not noteworthy on my resume. Keep it professional (Box 11.1 and 11.2).

Box 11.1 Training defined.

Going forward here, the term *training* in general is inclusive of formal education, technical training, and on-the-job experience.

Food Safety Lessons for Cannabis-Infused Edibles. https://doi.org/10.1016/B978-0-12-819512-3.00011-0

Because there is rapid growth in the cannabis industry, there are many applicants for jobs from people trying to get in but most of the applicants are unskilled and underqualified. The cannabis industry needs employees qualified for the jobs in cultivation and as trimmers, extraction technicians, and budtenders. Once an employee is trained in his/her job and if that person is a reliable employee, there will be opportunities to learn more skills on the job. Reliable employees will advance into promotions in a company and further their career. Unreliable employees will be terminated. Either way, the result is great turnover as employees advance or leave and companies grow; more qualified employees are needed. Companies in rapid growth are not able to spend enough time on training, continuous improvement, or development of cross-functional teams. From my time in the cannabis industry, I have observed companies using employees and teams who are dedicated to specific, limited tasks, e.g., a harvesting team. The employee and team become highly skilled in that task but are not trained for an additional role. However, the beauty of the employee dedicated to a task is the reduction of foot traffic and prevention of cross-contamination or cross-contact, as the employee has limited movement throughout the facility. I really like the team concept of employees being dedicated to a role, if the company can manage that.

The work of quality management is never done and in continuous improvement into perpetuity, but in the future, training will be a bit easier when procedures from starting of clippings to gas chromatography have been written to capture the knowledge of experienced employees, peer-reviewed research publications, validated methods, and industry best practices. Once standard methods are published and standard operating procedures (SOPs) as best practices are available, the documents will be widely used for training. In the meantime, this knowledge is held within individual companies. Some companies are good at writing documents for training, and some companies have no training documents. I am guessing our friends in cannabis laboratories have done a great job of writing training documents because the employees probably came from industries that required written, validated methods and SOPs. Even within cannabis laboratories, there is a closely held knowledge base of how to work with cannabis and its complications from waxy substances.

Training occurs for onboarding of temporary or new employees, refresher training, training to a new role or task, and corrective action on an as-needed basis. In an ideal world, supervisors would document at some frequency an observation of an employee performing a task. Training documentation should be stored in each employee's personnel file to demonstrate employee development and correlate to the employee's pay rate. The personnel file should hold transcripts of education, copies of certificates, proof of attendance at workshops or other professional development events, and sign-in sheets at company meetings where training is delivered. In industry, personnel files are managed by human resources. Some cannabis companies do not have a human resource professional and the work of documentation is put on the hiring manager. The National Cannabis Industry Association has published an Employee Onboarding Checklist for Managers as a tool for the hiring manager [1]. Feel free to steal this public document or use it as a guide to create your own. Regardless of who is responsible for maintaining personnel files or what the method is to accomplish the documentation of training, it must be done. Documenting training now will prepare a cannabis company for the future when FDA inspectors will review personnel files of any individual employee for evidence of that person being qualified to do his/her job.

Box 11.2 Myth: Cannabis employees get free weed.

States have very strict rules on compensating employees with cannabis product. Employees purchase from a dispensary like everyone else.

Resources specific to the cannabis industry

Along my journey, I have discovered valuable resources readily available for the cannabis industry. Across the country, there are workshops, course offerings, certificate programs, and degree programs. Awarding the first PhD with cannabis as the focus of research is claimed by the University of Guelph in Canada in 2018 [2]. A PhD from cannabis research will only come from countries where cannabis is legal; this leaves the United States in the dust. The United States is falling behind in cannabis research because public universities are federally funded and unable to conduct research. With marijuana as a Schedule I drug on the Drug Enforcement Administration list, universities are in fear of losing federal funding by growing cannabis or researching an illegal drug.

Cannabis is a new, dynamic, and rapidly growing field. By the time this book is published, much more will be available in the areas of reading materials, events for networking and offerings of cannabis workshops, courses, certificates, and degrees.

Reading materials

Read and read some more. There is so much information available publicly. Here is my reading list.

Anything from the FDA

You don't have to agree with the FDA, but it is nice to understand the FDA. Realize that the current thinking on cannabis, cannabidiol (CBD), and Δ^9-tetrahydrocannabinol (THC) is changing rapidly in the FDA, so older statements may not reflect the current thinking of the FDA. Try to read everything published by the FDA going forward on marijuana and cannabis, but start with these documents:

- Warning Letters and Test Results for Cannabidiol-Related Products [3].
- FDA and Marijuana, June 19, 2018 [4].
- Statement by FDA Commissioner Scott Gottlieb, MD, on the importance of conducting proper research to prove safe and effective medical uses for the active chemicals in marijuana and its components, June 24, 2018 [5].

- Statement from FDA Commissioner Scott Gottlieb, MD, on signing of the Agriculture Improvement Act and the agency's regulation of products containing cannabis and cannabis-derived compounds, December 20, 2018 [6].
- The FDA Commissioner testified on March 28, 2019, stating, "*We're using enforcement discretion right now ... I will take enforcement action against CBD products that are on the market if manufacturers are making what I consider over-the-line claims.*" Rather than citing the minutes of the hearing, which are available, the reference reports on the testimony. You're welcome [7].
- Statement from FDA Commissioner Scott Gottlieb, MD, on new steps to advance the agency's continued evaluation of potential regulatory pathways for cannabis-containing and cannabis-derived products, April 2, 2019 [8].
- FDA Regulation of Cannabis and Cannabis-Derived Products: Questions and Answers, April 2, 2019 [9].
- Scientific Data and Information about Products Containing Cannabis or Cannabis-Derived Compounds; Public Hearing, May 31, 2019 [10].
- What You Need to Know (And What We're Working to Find Out) About Products Containing Cannabis or Cannabis-derived Compounds, Including CBD, July 17, 2019 [11].
- FDA warns company marketing unapproved cannabidiol products with unsubstantiated claims to treat cancer, Alzheimer disease, opioid withdrawal, pain, and pet anxiety, July 23, 2019 [12].

Miscellaneous reading materials

- Anything by Ed Rosenthal on growing
- Search YouTube for an inside look into cannabis operations
- Textbooks recommended by any of the colleges or universities mentioned later
- Cannabis Industry Journal at www.cannabisindustryjournal.com
- Leafly Cannabis 101 [13]
- Cannabis-infused edibles cookbooks from your local library or available online
- Your state code for marijuana regulations or code from other states
- National Cannabis Industry Association [14]
- National Environmental Health Association: Cannabis Resources [15]

Events for networking

One of the best ways to learn about the cannabis industry is to attend cannabis meetings or conferences. If you are in a state with legal medical marijuana, there is a good chance of a meeting or conference in your state. If you don't want to spend a chunk of money on travel and you are in an area with a strong cannabis market, there will be plenty of local events sponsored by cannabis businesses. Networking might be as simple as listening to live music outside your local dispensary. Even at local events, there may be people from all over the United States and potentially other countries. Canada is very active in hosting conferences too. Listen to the presentations; talk to people in the lunch line or in the exhibit hall. There are many conferences aimed at different sectors of the cannabis industry covering laboratories, transportation, investment, and the state regulations. I found many conferences by searching for cannabis event calendars. Here is my short list of events for networking:

- Cannabis Quality Conference & Expo [16];
- Institute of Food Technologists, Legalized Cannabis and Hemp Edibles Group [17];
- Women Grow [18];
- National Cannabis Industry Association, quarterly caucuses and annual conferences [14].

Cannabis courses, certificates, workshops and webinars, and degrees

The focus of training here is on the individual employee in the cannabis industry and the person who wants to put cannabis training and education on his/her resume. The latter person may already be working in the cannabis industry and may want professional development or may not be working in the cannabis industry and want to apply his/her skillset from an aligned industry to the cannabis industry as I have done. This person could be a traditional student earning an undergraduate or graduate degree and be geeked out over learning about cannabis. There are training and education offerings to meet the needs of many, but not all.

Think of the following information as a start to a gap assessment of training and education by capturing what is available in a snapshot. The cited offerings are examples and are in no way exhaustive of the training available to the cannabis industry.

Courses

Cannabis courses are offered by colleges and universities within existing departments and areas of study, e.g., a horticulture department may offer a course on growing cannabis. There are courses in cannabis for lawyers, pharmacists, medical doctors, and business professionals. A student typically earns a grade for a course, which is documented on the student's transcript, and the transcript becomes part of the employee's personnel file.

Certificates

Certificates for courses or programs are offered by colleges and cannabis-specific organizations. The certificate may represent completing a series of courses in which the student earns a program certificate. Typically, a technical or community college will offer a certificate rather than a degree for a program that is completed in 1 year or less. In contrast, associate degrees are typically earned in 2-year programs.

Workshops and webinars

Starting in 1992, I worked as a full-time instructor in the Wisconsin Technical College System. At Northeast Wisconsin Technical College in Green Bay, I was the lead instructor for a 2-year associate degree, preparing students to work in a food or environmental laboratory. I taught 11 courses each year and I loved it! I am a huge proponent of adult education at the technical or community college and in the workshops and webinars I now deliver to industry. The difference between a course and workshop is there is no formal evaluation or grade given at a workshop. Workshops are most often taught in contiguous days, unlike courses taught in a semester. I have taught half-day workshops and 5-day workshops, but a typical workshop is 2 days. At the end, the workshop participant receives a certificate of attendance or participation. Workshops can be delivered in the format of webinars, with the advantage of not requiring the time and cost of travel, as the information is delivered remotely through a meeting platform. I have delivered 1-h webinars and webinars for one-half day on two consecutive days for a total of 7 h of instruction. Webinars are very well received as a form of training, and interactive platforms make learning effective.

Offerings for degrees, certificates, workshops, and webinars

The following are examples and do not constitute an exhaustive list. The mention of the offerings should in no way be interpreted as an endorsement.

- AIB Food Safety Essentials for Cannabis Producers and CBD Products [19].

  AIB offers an online Good Manufacturing Practices (GMPs) course.

- Allay Consulting [20].

Kimberly Stuck delivers an online GMP course in 2 hours. The course is available on Safe Food Training Hub.

- Cannabis Trainers [21].

Maureen McNamara teaches ServSafe to edible manufacturers, which is the requirement for food safety training for restaurants and minimum training required by some states for cannabis-infused edibles manufacturers and a program called SellSMaRT for dispensaries.

- Cannabis Training University [22].

  The Cannabis Training University offers 100% online certification programs.

- CBD Training Academy [23].

  CBD Training Academy offers live or on-demand webinars.

- Oaksterdam University [24].

Oaksterdam is dedicated to cannabis in Oakland, California, with the emphasis on horticulture and business. There is an online option.

- Clover Leaf University [25].

Course descriptions are found in the catalog and include Cannabis Science and Testing, Responsible Manufacturer (Infused Product) Certification, Cannabis Culinary Certification, and Infused Laboratories - Cannabis Extractions.

- THC University [26].

  The THC University offers 100% online certification programs.

- HEMP STAFF [27].

  HEMP STAFF offers budtender training across the United States.

- Trichome Institute [28].

The Trichome Institute offers cannabis products and sales training (CPST) certification for budtenders and the unique offering of a cannabis "sommelier" course called Interpening.

University degrees

- The Dalhousie University in Canada offers eight modules on growing operations that are 100% online [29].

Undergraduate degrees in the United States in medicinal plant chemistry

- Northern Michigan University [30].
- Minot State University, North Dakota [31].

Graduate degrees in Canada

With cannabis legalized in Canada, Canadian universities can do cannabis research and offer both undergraduate and graduate degrees specifically in cannabis.

- Loyalist College [32].

Entrance into the Cannabis Applied Science graduate program requires strong math and science skills with an undergraduate degree from either Loyalist College in BIOS/ENVR/CHEM or another college or university, with the completion of eight semesters of theory courses and six semesters of laboratory courses in chemistry and biology. Students earn a One-year Ontario College Graduate Certificate.

- Guelph Centre for Cannabis Research at the University of Guelph [33].
- McGill University [34].
- McMaster University [35].

Resources in related fields

Are you in the food industry? What training did you need for your position? The needs of cannabis-infused edibles manufacturers align with the needs of the food industry. They need people with expertise in logistics, IT, facility management, production scheduling, procurement, accounting, marketing, sales, human resources, and maintenance. A cannabis-infused edible manufacturer will need food scientists specializing in equipment engineering, production, sanitation, sensory science, quality assurance, food safety, regulatory affairs, packaging engineering, and labeling. The food industry relies on a family of industries in the farm-to-fork continuum, with the food industry sandwiched between agriculture and retail sales. Members of the food industry work for government, industry, academia, or associations. The same is true for the cannabis industry. Unique to cannabis is a heightened need for employee, ingredient, and product security. Take what the food industry has done for food defense and multiply by 10 just to start for documentation, monitoring, and secure access, then add a budget for guards and armored vehicles. Like the food industry, the third-party cannabis laboratory operates independently and off-site. Depending on a requirement for vertical integration, a growing operation and extraction facility may be on- or off-site from the edibles manufacturer's site.

I hope the cannabis industry is learning from the aligned industries of food, dietary supplements, pharmaceutical, medical devices, and tobacco. Let me remind you that these industries are regulated under the umbrella of the FDA at the federal level. From my perspective of food safety, I am familiar with the training needed by the food industry and resources for that training. Certainly, the cannabis industry can use the same food safety training and apply the knowledge to the manufacturing of edibles, until cannabis-specific training is developed. Check back with me later in 2020 on that; I have exciting adventures in cannabis training in the hopper. In addition to training, the cannabis industry and edibles manufacturers are taking from the food industry the use of its equipment, pest control management, cleaning and sanitizing procedures, and laboratory methods. Each of those applications requires site-specific training as well.

Training needed for personnel in cannabis-infused edible manufacturing

The following list focuses on the training needs to achieve edible safety as a good start but not an exhaustive list. Personnel in cannabis-infused edible manufacturing need training on

1) GMPs for every employee,
2) allergen management for every employee,
3) SOPs and policies for visitors and contractors,
4) food defense and facility security for every employee
   a. Food Safety Preventive Controls Alliance (FSPCA) free online training for employees and supervisors,
5) certification of one onsite Preventive Controls Qualified Individual to meet the FDA requirement,
6) certification of at least one additional Preventive Controls Qualified Individual as a backup person,
7) SQF practitioner or similar certification to meet auditing standard, optional,

8) SOPs specific to job duties for every employee
   a. sanitation SOPs for the sanitation crew,
9) overview of state code and FDA Preventive Controls for Human Food (PCHF) rule for management,
10) food safety culture for management and supervisors.

I can only speak to my own experience of delivering and receiving training in the food industry. As an educator, I have affiliations with some of the following organizations, but not all. These are the leaders I would point any food scientist or cannabis professional to for information and training.

General resources for personnel in cannabis-infused edible manufacturing

- The American Society for Quality (ASQ) offers certification for quality professionals [36].
- EAS Consulting Group has on-demand public webinars on cannabis, dietary supplements, and food [37].
- Ecolab is a leader in cleaning and sanitizing for the food industry [38].
- Food Allergy Research and Resource Program (FARRP) [39].
- Food Protection and Defense Institute (FPDI) [40].
- FSPCA for training on the FDA Food Safety Modernization Act (FSMA) rules:
  - Foreign Supplier Verification Program rule [41].
  - Intentional Adulteration rule [42].
  - Preventive Controls for Animal Food rule [43].
  - PCHF rule [44].
- Global Food Safety Initiative (GFSI)-recognized certification programs [45].
- The food industry in the United States is dominated by SQF, a GFSI-recognized program [46].
- ImEpik has a 100% online course to earn a certificate for Preventive Controls Qualified Individual [47].
- Institute of Food Technologists [48].
- International Association for Food Protection [49].
- NSF International offers private, public, and webinar training:
  - Canada food courses [50].
  - US food courses [51].
  - Dietary supplements training [52].
- SGS Group is an international organization providing auditing and other services [53].
- Third-party laboratories will work with clients and some offer workshops.
- University extension specialists in food safety or horticulture.

A final thought

ServSafe and Hazard Analysis and Critical Control Points (HACCP) training are not appropriate for most cannabis-infused edibles manufacturers. The only reason a cannabis-infused edibles manufacturer should send employees to ServSafe training is if the state code requires all employees to be ServSafe trained and the employees are entry level and not supervisors. All employees should be trained in GMPs; ServSafe covers only a small part of the GMP material.

HACCP started in the food industry in the early 1960s and has been replaced by the FDA rule for PCHF. The food industry under FDA has left HACCP as controlling hazards and moved on to PCHF. Why would the cannabis industry be satisfied with an old concept that was not good enough for the food industry?

At some point in the history of HACCP in the food industry, the restaurant industry adopted the steps of HACCP to control hazards in a restaurant. As a frequent patron to restaurants, I applaud the restaurant industry for their control of hazards through HACCP, ServSafe, and the Food Code. Cannabis-infused edible manufacturers are not restaurants and it is inappropriate to hold the manufacturers to the same standard as a restaurant. HACCP, ServSafe, and the Food Code are a good starting point to understand food safety, but there is so much more. By implementing the requirements of FDA's FSMA rule for PCHF, the cannabis industry is up to the same level as the food industry and, more importantly, can strive to control hazards in edibles and protect the consumer.

References

[1] National Cannabis Industry Association. Employee onboarding checklist for managers. https://thecannabisindustry.org/wp-content/uploads/2019/08/HRC_OnBoarding_Managers-Checklist_July.pdf. [Accessed 09.01.19].

[2] The Star. Cannabis PhD takes higher education to a new level. https://www.thestar.com/news/cannabis/2018/09/05/cannabis-phd-takes-higher-education-to-a-new-level.html. [Accessed 09.01.19].

[3] Food and Drug Administration. Warning letters and test results for cannabidiol-related products. https://www.fda.gov/news-events/public-health-focus/warning-letters-and-test-results-cannabidiol-related-products. Content current as of: 07/24/2019. [Accessed 09.14.19].

[4] Food and Drug Administration. FDA and marijuana. June 19, 2018. https://www.fda.gov/news-events/public-health-focus/fda-and-marijuana. [Accessed 09.01.19].

[5] Food and Drug Administration. FDA statement: statement by FDA Commissioner Scott Gottlieb, M.D., on the importance of conducting proper research to prove safe and effective medical uses for the active chemicals in marijuana and its components. June 24, 2018. https://www.fda.gov/news-events/press-announcements/statement-fda-commissioner-scott-gottlieb-md-importance-conducting-proper-research-prove-safe-and. [Accessed 09.01.19].

[6] Food and Drug Administration. FDA statement: statement from FDA Commissioner Scott Gottlieb, M.D., on signing of the Agriculture Improvement Act and the agency's regulation of products containing cannabis and cannabis-derived compounds. December 20, 2018. https://www.fda.gov/news-events/press-announcements/statement-fda-commissioner-scott-gottlieb-md-signing-agriculture-improvement-act-and-agencys. [Accessed 09.01.19].

[7] FDA chief clarifies enforcement priorities for CBD products. https://www.marijuanamoment.net/fda-chief-clarifies-enforcement-priorities-for-cbd-products/. [Accessed 09.01.19].
[8] Food and Drug Administration. FDA statement: statement from FDA Commissioner Scott Gottlieb, M.D., on new steps to advance agency's continued evaluation of potential regulatory pathways for cannabis-containing and cannabis-derived products. April 2, 2019. https://www.fda.gov/news-events/press-announcements/statement-fda-commissioner-scott-gottlieb-md-new-steps-advance-agencys-continued-evaluation. [Accessed 09.01.19].
[9] Food and Drug Administration. FDA regulation of cannabis and cannabis-derived products: questions and answers. April 2, 2019. https://www.fda.gov/news-events/public-health-focus/fda-regulation-cannabis-and-cannabis-derived-products-questions-and-answers. [Accessed 09.01.19].
[10] Food and Drug Administration. Public hearing: scientific data and information about products containing cannabis or cannabis-derived compounds. May 31, 2019. https://www.fda.gov/news-events/fda-meetings-conferences-and-workshops/scientific-data-and-information-about-products-containing-cannabis-or-cannabis-derived-compounds. [Accessed 09.01.19].
[11] Food and Drug Administration. Consumer update: what you need to know (and what we're working to find out) about products containing cannabis or cannabis-derived compounds, including CBD. July 17, 2019. https://www.fda.gov/consumers/consumer-updates/what-you-need-know-and-what-were-working-find-out-about-products-containing-cannabis-or-cannabis. [Accessed 09.01.19].
[12] Food and Drug Administration. FDA news release: FDA warns company marketing unapproved cannabidiol products with unsubstantiated claims to treat cancer, Alzheimer's disease, opioid withdrawal, pain and pet anxiety. July 23, 2019. https://www.fda.gov/news-events/press-announcements/fda-warns-company-marketing-unapproved-cannabidiol-products-unsubstantiated-claims-treat-cancer. [Accessed 09.01.19].
[13] Leafly. Cannabis 101. https://www.leafly.com/news/cannabis-101. [Accessed 09.01.19].
[14] National Cannabis Industry Association. https://thecannabisindustry.org/. [Accessed 09.01.19].
[15] National Environmental Health Association. Cannabis resources. https://www.neha.org/node/60143. [Accessed 09.01.19].
[16] Cannabis Quality Conference & Expo. https://cqcexpo.com/. [Accessed 09.01.19].
[17] Institute of Food Technologists. Legalized Cannabis and Hemp Edibles Group. https://connect.ift.org/groups/home/261. [Accessed 09.01.19].
[18] Women Grow. https://womengrow.com/. [Accessed 09.01.19].
[19] AIB International. Food safety essentials for cannabis producers and CBD products. https://www.aibinternational.com/en/Training/Food-Safety-Quality/GMP-Sanitation-Training/Online-Training/Food-Safety-Essentials-for-Cannabis-Producers. [Accessed 09.01.19].
[20] Allay Cannabis Consulting. Cannabis food safety training course. https://allayconsulting.com/cannabis-food-safety-fda-gmp-compliance-in-product-manufacturing-dispensaries-and-grows/. [Accessed 09.01.19].
[21] Cannabis Trainers. https://cannabistrainers.com/about-us/. [Accessed 09.01.19].

[22] Cannabis Training University. https://cannabistraininguniversity.com/. [Accessed 09.01.19].
[23] CBD Training Academy. https://cbdtrainingacademy.com/on-demand-courses/. [Accessed 09.01.19].
[24] Oaksterdam University. https://oaksterdamuniversity.com/academics/. [Accessed 09.01.19].
[25] Clover Leaf University. https://www.cloverleafuniversity.com/. [Accessed 09.01.19].
[26] THC University. https://thcuniversity.org/. [Accessed 09.01.19].
[27] HEMP STAFF. Upcoming cannabis dispensary training courses. https://www.hempstaff.com/cannabis-training-for-dispensary-jobs. [Accessed 09.01.2019].
[28] Trichome Institute. https://trichomeinstitute.com. [Accessed 09.01.19].
[29] Dalhousie University. Cannabis production & management. https://www.dal.ca/faculty/agriculture/extended-learning/Courses/cannabis-production—management.html. [Accessed 09.01.19].
[30] Northern Michigan University. Medicinal plant chemistry. https://www.nmu.edu/chemistry/medicinal-plant-chemistry. [Accessed 09.01.19].
[31] Minot State University. Medicinal plant chemistry. https://www.minotstateu.edu/enroll/programs/medicinal-plant-chemistry.shtml. [Accessed 09.01.19].
[32] Loyalist College. Cannabis applied science. https://www.loyalistcollege.com/programs-and-courses/full-time-programs/cannabis-applied-science/. [Accessed 09.01.19].
[33] University of Guelph, Guelph Centre for Cannabis Research. Globe and mail features work of cannabis production expert. https://news.uoguelph.ca/2019/05/globe-and-mail-features-work-of-cannabis-production-expert/. [Accessed 09.01.19].
[34] McGill University, Agricultural and Environmental Sciences. Cannabis PhD takes higher education to a new level. https://www.mcgill.ca/macdonald/channels/news/cannabis-phd-takes-higher-education-new-level-289458. [Accessed 09.01.19].
[35] McMaster University. The science of cannabis certificate program. https://cannabisresearch.mcmaster.ca/. [Accessed 09.01.19].
[36] ASQ, Certification. The value of an ASQ quality certification. https://asq.org/cert/right-for-you. [Accessed 09.02.19].
[37] EAS Consulting Group. On-demand webinars. https://easconsultinggroup.com/resources/on-demand-webinars/. [Accessed 09.02.19].
[38] Ecolab. Food & beverage processing. https://www.ecolab.com/about/industries-we-serve/food-and-beverage-processing. [Accessed 09.02.19].
[39] Food Allergy Research and Resource Program. https://farrp.unl.edu/. [Accessed 09.02.19].
[40] Food Protection and Defense Institute. https://foodprotection.umn.edu/. [Accessed 09.02.19].
[41] Food Safety Preventive Controls Alliance. Foreign supplier verification program. https://www.ifsh.iit.edu/fspca/courses/foreign-supplier-verification-programs-fsvp. [Accessed 09.02.19].
[42] Food Safety Preventive Controls Alliance. Intentional adulteration. https://www.ifsh.iit.edu/fspca/courses/intentional-adulteration. [Accessed 09.02.19].

[43] Food Safety Preventive Controls Alliance. Preventive controls for animal food. https://www.ifsh.iit.edu/fspca/fspca-preventive-controls-animal-food. [Accessed 09.02.19].

[44] Food Safety Preventive Controls Alliance. Preventive controls for human food. https://www.ifsh.iit.edu/fspca/fspca-preventive-controls-human-food. [Accessed 09.02.19].

[45] Global Food Safety Initiative (GFSI)-recognized certification programs. https://www.mygfsi.com/certification/recognised-certification-programmes.html. [Accessed 09.02.19].

[46] The food industry in the United States is more concentrated in SQF, a GFSI-recognized program and materials. https://www.sqfi.com/resource-center/sqf-guidance-documents-audit-checklists/. [Accessed 09.02.19].

[47] ImEpik. www.imepik.com. [Accessed 09.02.19].

[48] Institute of Food Technologists. www.ift.org. [Accessed 09.02.19].

[49] International Association for Food Protection. www.foodprotection.org. [Accessed 09.02.19].

[50] NSF International, Canada. Explore our courses. https://www.nsflearn.com/ca. [Accessed 09.02.19].

[51] NSF International, United States. Explore our courses. https://www.nsflearn.com/us. [Accessed 09.02.19].

[52] NSF International. Dietary supplements training and education. http://www.nsf.org/services/by-industry/nutritional-products/dietary-supplements-training. [Accessed 09.02.19].

[53] SGS Group. Agriculture and food. https://www.sgsgroup.us.com/en/agriculture-food. [Accessed 09.02.19].

CHAPTER

Jobs in the cannabis industry

12

Transfering skills to the cannabis industry

So do you want to transfer your skill set to the cannabis industry like I did? My journey is unique, and your journey will be unique. The skills I have transferred to the cannabis industry include Good Manufacturing Practices (GMPs) gap assessment; swabbing to perform a risk assessment for pathogens in the environment; writing Hazard Analysis and Critical Control Points (HACCP), food safety, allergen management, and food defense plans; and interpreting FDA regulations. I do a lot of writing and training. Your journey will not look like my journey any more than you and I had a similar journey in the food industry. Any job in the food industry can transfer to the cannabis-infused edibles industry. Like the food industry, the cannabis industry is both local and international, vast, and complex. You will find your niche in the cannabis industry.

Myth

If you work in the cannabis industry, you get high

No. I have had people working in the cannabis industry tell me that they never get high, usually in a whisper. Δ^9-Tetrahydrocannabinol (THC) did not enter my body until December 24, 2018. After consuming cannabis-infused chocolate, one time in 2019, I had my body relax (thank you *Cannabis indica*) without other psychoactive effects. Otherwise, I have never been high, and I work in the cannabis industry.

Myth

It's too late to get into the cannabis industry

Never! Is it too late to get into the food industry? No, there will always be a need for food professionals. The cannabis industry in the United States has barely begun. Not every state has legalized medical marijuana yet. The need for experienced employees is great and will only grow as more states legalize medical marijuana and recreational cannabis. The hemp industry is growing with the signing of the Farm Bill in 2018, and CBD products are everywhere. When marijuana is descheduled, there will be an explosion of opportunities in government, industry, and academia.

Food Safety Lessons for Cannabis-Infused Edibles. https://doi.org/10.1016/B978-0-12-819512-3.00012-2

Where do I start for a job in the cannabis industry?

Learning about the cannabis industry looks like training for the cannabis industry. Follow the opportunities for education and training as presented in Chapter 11, Training Needed in the Cannabis Industry. Many of the resources are free, and your journey in the cannabis industry can start for free with reading.

Read

Reading includes watching YouTube videos and listening to on-demand webinars and patient testimonials. Good job reading this book too! Prepare to go down a rabbit hole for hours as you immerse yourself in reading. Find out whether your state has legal medical marijuana and recreational use and then find the state code for marijuana regulations. Start with just your state; understand your state. If your state, like my state of Wisconsin, does not have either medical or recreational, then research the hemp industry in your state or pick a neighboring state that has legalized recreational use and find that state's regulations for both medical marijuana and recreational use. If you are located far from legal recreational use cannabis like Hawaii or Key West, Florida, just pick a state and read the code. Read the Canadian cannabis regulation. Folks in Hawaii and Florida may still choose to read their state's code for medical marijuana, which is a big business in Florida.

Field trips

Create an adventure for yourself by visiting a dispensary, which I call a field trip. To shop in states that only have medical marijuana requires a state-issued medical card. Without a medical card, you cannot enter a dispensary in those states. Folks from Hawaii have to fly to shop at a dispensary so that's a tough spot, and folks from Key West have to drive hundreds of miles or also fly to shop at a dispensary. If you are lucky enough to live near a dispensary in a state with recreational cannabis, go! As I travel to states with recreational cannabis, I try to make time for a field trip. Most dispensaries retrofit into existing buildings, so the variation is wide. Every dispensary I have visited has been unique in the same way that every food facility is different and memorable.

In a state like Wisconsin where there are no dispensaries, we have cannabidiol (CBD) stores. Anyone can enter a CBD store. Give yourself a field trip to the local CBD store to learn about the products, packaging, and labeling. Ask a lot of questions. Good CBD stores and their employees are prepared to educate the consumer.

Network

Attend cannabis meetings and conferences hosted locally near you or plan to attend a large conference in state or out of state. I was lucky enough in 2018 to attend the National Cannabis Industry Association (NCIA) annual conference in San Jose when I piggybacked my travel to deliver food safety training at a food facility nearby. NCIA conferences cover seed to sale to learn about the whole industry, and you can find cannabis conferences specific to the interests of growing, business management, or medical research.

Depending on where you live or where you travel, there may be cannabis events. You can find these in the usual way you find any local event. You can get good information by searching for cannabis event calendars or finding websites that post cannabis events, such as Oaksterdam University in Oakland, California [1].

If you find someone with a medical card then have them describe their experiences with purchasing products and relief of symptoms of their condition. My cousin in Minnesota shared with me that he talked to a fellow tenant with multiple sclerosis about her condition and the use of marijuana. I met a young woman in the lunch line at a cannabis conference in 2018 who lost her father on 9/11, as he worked for the Port Authority of New York & New Jersey and died in one of the towers. Years of counseling and prescription drugs could not relieve her post-traumatic stress disorder. It was only after the use of marijuana that she started to feel functional again. Every patient has a story to tell.

Know the jobs

Many cannabis websites include job boards. Browsing through job postings is a good way to learn the roles of cannabis employees and their job descriptions. You can find a job that interests you and prepare for that future job or you can align your current skill set to a current job posting and their job description and apply! Here are a few sites to start your job search.

Vangst is dedicated to the cannabis industry for job postings, a database of employees, and gigs. I am relatively new to the cannabis industry, but I love the Vangst website. It was on the Vangst website that I first saw a job posting for food scientist at a cannabis company (March 15, 2019); my heart melted. Here are excerpts of a QA manager job posted by Vangst [2].

> *Reporting to the Director of Operations for Extraction and Production, QA & Regulatory Manager will steward critically important, high visibility projects across the division. The QA & Regulatory Manager is responsible working with the Principal Engineer to drive/maintain processes and configuration management practices to bring the division into alignment with cGMP (21 CFR) as well as oversee compliance with the reporting requirements set forth by the State of Nevada (METRC). They will be principally involved with the development of a total Quality Management System (tQMS) and training staff across the division in best practices to facilitate total traceability and transparency. To ensure that products are absolutely fit for intended use, the QA & Regulatory Manager will see to it that the division can demonstrate in a documented form that the processes, methods, tests, activities and equipment's deployed are capable of repeatedly producing the desired product.*

Just a few of the 14 listed Requirements for the QA Manager:

1. *Bachelor's Degree (Master's a plus) in a relevant discipline (e.g., chemistry, biological sciences, etc.) and preferably holding Quality Assurance Professional (ASQ) Certification.*
2. *Knowledge of Annex 1, Annex 16, FDA, EMA, Health Canada, MHLW, and ICH regulatory requirements and other guidelines related to cGMP operations.*
3. *Minimum of 3–5 years' experience working at the supervisory level in regulated industry (pharmaceutical, cannabis, alcohol ...) preferably cannabis related.*
4. *Experience performing audits against domestic and international regulatory requirements and industry expectations as related to cGMP operations.*
5. *Demonstrated experience generating the totality of documentation required for GMP certification and compliance.*

HEMP STAFF offers training across the United States for budtenders and posts a job board [3].

Many of the training organizations discussed in Chapter 11 post a job board. Those go hand in hand, I guess. Oaksterdam is dedicated to training in cannabis horticulture and business and posts a job board [4].

Availability of jobs in the cannabis industry

What jobs are available in the cannabis industry? Continuing from above on the search for jobs and job descriptions on job boards, this article in Business Student does a nice job of detailing 26 job titles in the cannabis industry [5]. A way to look at job sectors is to see the state code for licensing. In Colorado, there is a license for each of growing, processing, and manufacturing, transportation, laboratory, and dispensary. By focusing on the sector of seed to sale where you want to work, you can focus on the jobs available. There are jobs in every sector of the cannabis industry:

- Growing Operations
- Processing, including Extraction
- Manufacture of Edibles and Topicals
- Transportation
- Cannabis Laboratory
- Dispensary

There are three levels of employees needed in the cannabis industry: (1) entry-level employees, (2) mid-management supervisors, and (3) management. Upper management is composed of investors, owner, a business management group akin to "Corporate," and the facility manager similar to a plant manager in the food industry. Mid-management positions include department managers who are direct reports to the facility manager. If the facility employs a large number of people, there may be supervisory positions reporting to department managers. Supervisors will be needed for each department or function on-site at the facility. Entry-level employees are hands-on. They include employees who handle and care for plants, employees who harvest flower, employees who trim dried flower, the sanitation crew, the ingredient weigh-up technicians, and drivers. Entry-level employees are highly specialized, usually with manual labor, and require on-the-job training to master their job. Managers, on the other hand, are relatively few in a cannabis facility and wear many hats to cover all the departmental requirements and responsibilities. A quality assurance manager may have responsibilities in employee training; document management; regulatory affairs; compliance with state code for sampling, packaging, and labeling; hold-and-test programs for ingredients and products; auditing; and sanitation. Quality assurance managers know what I mean here; they have many responsibilities that overlap into many departments.

The job you get in the cannabis industry will depend on your skill set and your level of experience. There are tons of entry-level jobs in growing operations and dispensaries. If you are looking to just get into the cannabis industry, you can start at either end of growing or selling in a dispensary, depending on what is more attractive to you. There are fewer jobs in the processing of cannabis or hemp with extraction and edibles manufacturing. To work hands-on in extraction or in a laboratory, you should be an experienced chemist. Understanding organic and analytical chemistry is your ticket to get a job in extraction or in chemical analysis in a cannabis laboratory. Cannabis laboratories are also in need of applicants with experience in microbiology or data entry. If you are coming to the cannabis industry with hands-on food production experience in a restaurant or food facility, your resume will be attractive to an edible manufacturer.

Departmental needs within a cannabis-infused edibles manufacturer

A cannabis-infused edibles manufacturer will need employees, contractors, or consultants with expertise in a multitude of departments and functions. At a start-up facility, employees will wear many hats to get the job done or the company will contract the work out on a project basis, until employees with expertise can be hired. For example, I imagine IT work can be contracted to a service company. Taking a cue from the food industry, I see the following departmental needs specific to all cannabis-infused edibles manufacturers.

- Product development in the form of a chef or food scientist
- Regulatory affairs to include packaging and labeling law
- Purchasing
- Management and administration to include human resources and accounting
- Facility operation and maintenance to include engineering
- Quality assurance to include sensory analysis, regulatory compliance, and auditing
- Facility security to include food defense
- Production from batching of ingredients to packaging and labeling
- Sanitation
- Transportation specific to receiving cannabis ingredients and shipping cannabis product
- Receiving
- Storage on-site
- Shipping of product

I probably missed a major department, but whatever I missed can go under one of the listed departments such as management. I expect most cannabis companies do not have a human resources department, and hiring managers handle the legal requirements of hiring and firing. Where would you put OSHA compliance? It could be addressed by management, facility operations, or regulatory affairs. Large companies may have some of the departments off-site at "Corporate," e.g., regulatory affairs and purchasing can be consolidated for multiple facilities and be off-site. Depending on the state code, product development may be required to be on-site at an edible facility because of security around the cannabis ingredient. The state code may also have requirements for product sampling procedures to be done by cannabis laboratory technicians and not edible facility employees. Know the code.

The food industry can learn from the cannabis industry

The food industry takes federal oversight for granted. The cannabis industry is *fighting* for descheduling of marijuana off the Drug Enforcement Administration list and for reasonable regulations. The cannabis industry is passionate about its causes.

Along with the passion in the cannabis industry comes the entrepreneurial spirit. Almost everyone in the cannabis industry is fresh and new, full of hope, and determined. Entrepreneurs abound in all sectors of the industry. They are taking chances with their energy, time, and funds. They are taking risks.

The cannabis industry lives and dies by the state code. Food lawyers and employees in regulatory affairs take the time to read regulation in the food industry. When I read the updated Canadian code for the Cannabis Act that addressed edibles, I was very impressed with the simple language. I think more people in the food industry should know how to locate rules and guidance documents and read those to be familiar with the contents and language.

Finally, the cannabis industry is by far winning in the use of personal protective equipment (PPE) compared to the food industry. Employees handling plants, dried plant material, and edibles are covered in PPE. The food industry could do better in enforcing the use of PPE and requiring full coverage when employees are handling ready-to-eat products.

Characteristics of a cannabis industry employee

From my observations of employees, there is a heightened attention to detail in each position. There is a lot of work to just meet the regulation requirements, and every employee must know his/her job. Because many companies are start-up businesses, managers must wear many hats and cover many functions at the facility. Other characteristics of a cannabis industry employee include flexibility, integrity, driven to work hard, and readiness for promotion.

Flexibility

If you are a quality manager who likes the predictability of the rules and uses the rules as benchmarks, you may not like the cannabis industry. The rules, i.e., state code and regulations, are changing all the time. For example, in 2018, I heard the state of Washington took the production of one type of edible out of regulation, which caused a relatively large manufacturer to shut down a whole line of product. Then on August 1, 2019, Washington released a statement [6] that the state would align with the FDA stance on hemp-derived CBD in food and beverage products, meaning the CBD products are illegal, and the state would work with industry as product is removed from the market. The new rule took effect on … August 1, 2019, the same day, and stated:

> *Other parts of the hemp plant, including CBD, cannot be used as a food ingredient under a Washington State Food Processor License [7].*

If you have the drive to optimistically roll with the punches and strive to meet challenges come on board for the cannabis ride! The ride is going to be bumpier when marijuana is descheduled and FDA has oversight of edibles and other products. With professionals in the cannabis industry, the industry can drive regulation with what is already working before the time comes for FDA regulation.

Integrity

First for the employee, this is more than passing the mandatory background check for employment. Integrity is doing the right thing even when nobody is watching. There are cameras everywhere in a cannabis facility, and people are watching. If you are an employee in the cannabis industry, you can't even think of slipping a bit of product in your pocket. Not only will the act of stealing be seen on a camera but also the act will probably be recorded from multiple angles. Like employee stores at a food facility that sell product at a reduced rate, cannabis employees have stores too. They are company-owned dispensaries, and employees get a small discount. Otherwise, the sale to an employee is handled exactly like a non-employee visitor to the dispensary. Second, management integrity is critical for the long-term success of the company. Management must set the culture of the facility by following the law. It is illegal for management to pay or incentivize a cannabis employee with product.

Driven to work hard

Entry-level jobs require manual labor or being on your feet for an extended time, but there must also be a mental drive within a cannabis employee. In addition to mastering their job and attention to detail, a great employee will be educated in cannabis. They will be passionate about cannabis in their own way and learn everything they can on and off the job. Successful employees will show continuous improvement and willingness to learn. If not, they will move along to a different industry.

Readiness for promotion

The result of a cannabis employee having flexibility, integrity, and drive to work hard is that the employee will be eager to learn more responsibilities and be promoted. As a cannabis company is successful and grows, successful employees will change departments or grow into newly created positions. This process does not take a great deal of time. When I ask cannabis employees how long they have worked at a company, I often get, *A long time. A little over a year.* I have included the feature of Real People in Cannabis in this book to show a variety of people have come into the cannabis industry and built careers in a relatively short amount of time. The cannabis industry will be born when marijuana is descheduled. In the meantime, the cannabis industry is in gestation and preparing for its birth. Like the prenatal growth of a baby, important systems are synthesizing and milestones are being achieved. My journey in the cannabis industry started in late 2017 with a phone call with Rick Biros and his son, Aaron Biros, of Innovative Publishing. I wrote a series of articles on microbiology and HACCP for Aaron's online publication, *Cannabis Industry Journal*. Now I have a cannabis book, and I am still learning immense amounts of information for application in the cannabis industry (Box 12.1).

Box 12.1 Real People in Cannabis

Drew Towns, Director of Operations, Sozo Extracts, Louisville, KY.

Time in the cannabis industry: Worked with Sozo for 4 years in cannabis/hemp space.

What drives me in my job: The cannabis plant has been shrouded in secrecy and propaganda for over 80 years. I feel lucky to be alive during a time where we get to eradicate negative perception and build a new industry from the ground up.

References

[1] Oaksterdam University: Events. https://oaksterdamuniversity.com/events/. [Accessed 09.15.19].

[2] QA Manager. https://www.vangst.com/job-details?id=1479. [Accessed 09.09.19].

[3] HEMP STAFF. https://www.hempstaff.com. [Accessed 09.03.19].

[4] Oaksterdam University. https://oaksterdamuniversity.com/employment-listings/. [Accessed 09.03.19].

[5] Steele HR. 25+ best marijuana cannabis jobs & careers, business student. December 9, 2017. https://www.businessstudent.com/careers/cannabis-industry-careers/. [Accessed 09.03.19].

[6] Leafly: Washington state bans CBD in food and beverages. https://www.leafly.com/news/politics/washington-state-bans-cbd-in-food-and-beverages. [Accessed 09.03.19].

[7] State of Washington, Department of Agriculture. Restrictions on the use of hemp CBD as a food ingredient. August 1, 2019. https://cms.agr.wa.gov/getmedia/595b6e0a-8490-40cf-8fd6-7669f1610c60/CBD_Statement_08-01-19-(002). [Accessed 09.03.19].

CHAPTER

Final thoughts

13

Marijuana Business Conference (MJBizCon)

I attended MJBizCon December 10–13, 2019, in Las Vegas, Nevada. I believe the conference assertion that MJBizCon is the largest cannabis business conference in the world based on their data of 33,000 attendees and 1300 exhibitors at the Las Vegas Convention Center [1]. To navigate the craziness, the conference created an excellent, free app with the schedule, speaker bios, exhibitor descriptions, and the capabilities to create a personal schedule and network with attendees (Fig. 13.1).

FIGURE 13.1

MJBizCon attendee badge.

Personal photo of the author.

Food Safety Lessons for Cannabis-Infused Edibles. https://doi.org/10.1016/B978-0-12-819512-3.00013-4

Cannabis associations

This year was the first year of MJBizCon *week*, starting with workshops on Monday and Tuesday. I flew into Las Vegas too late to attend the launch of MJBizCon Associations Day, unfortunately, on Tuesday. Kudos to the conference organizers for bringing together and celebrating 26 associations! I had no idea there were that many cannabis associations working on behalf of the cannabis industry. Some of the associations are more narrowly focused in their work, e.g., the Missouri Medical Cannabis Trade Association, but most work broadly across the nation, e.g., the Marijuana Policy Project, and internationally. I now realize there are many state associations within most states working on behalf of the cannabis industry. My quick Google search found the Wisconsin Medical Marijuana Association and Wisconsin Cannabis Trade Association in my home state. I also found prolific writing on the intersection of cannabis and Wisconsin on the website CannaBadger.com. I love the name CannaBadger, incorporating my beloved mascot of the Univeristy of Wisconsin-Madison, Bucky Badger.

Presentations at MJBizCon

I thoroughly enjoyed the MJBizCon presentations to kick off Wednesday starting at 9:00 a.m. I heard a welcome; state of the cannabis industry; keynote address by Tom Daschle, former senator from South Dakota; and a panel discussion of the 2020 outlook. I appreciated that the first day of presentations was a single track compared with the second and third days having up to five tracks of presentations from which to pick. The available information was overwhelming, and attendees were told that we would have access to PowerPoint presentations in the following week. Key learnings from the first day of presentations included the following:

- MJBizCon focuses on the business aspects of the cannabis industry, including vendor sourcing, entrepreneurialship, investor education, and regulatory affairs.
- MJBizCon is a global event with attendees from 79 different countries [1]. Attendees came from Canada, Mexico, Central America, South America, the United Kingdom, Europe, Israel, Northern Africa, South Africa, Russia, India, China, Indonesia, Australia, and New Zealand.
- Illinois will start the sales of recreational cannabis on January 1, 2020, and has the potential for a $2 billion market at maturity.
- California companies are being heavily taxed. It is going to get worse, before it gets better. The year 2019 will see a $1 billion growth in revenue from 2018.
- In general, there is a merger and acquisition slow down.
- The trend of peer companies buying peer companies is changing to larger companies buying smaller companies and noncannabis companies merging with cannabis companies in the same sector. For example, AriZona Beverage Company, maker of AriZona Iced Tea, partnered with Dixie Brands for a licensing agreement. Dixie Brands of Colorado produces the cannabis-infused beverage, Dixie Elixirs.
- Oklahoma has a booming hemp business because of the industry being loosely regulated compared with other states. Licenses were first available in Oklahoma in July 2018. In 2020, I later learned there are 1500 licenses for dispensaries in Oklahoma.
- New York was heavily represented at MJBizCon.

- Because the Farm Bill legalized the growing of hemp, there is good opportunity for cannabidiol (CBD)-infused pet care products and supplements.
- There is a CBD boom as sales of CBD products go into large-chain retail stores.
- The vape crisis originated in the illegal black market.
- As vape sales decreased with the vape crisis, sales of cannabis-infused edibles increased.
- MJBizCon started with 400 attendees and 4500 attendees just 4 years ago.
- Illinois has a model recreational cannabis bill to include social equity and can be used for what works and what doesn't work as other states write social equity into their cannabis bills.
- FDA has delayed guidance on CBD products, citing lack of data.
- Canadian stocks tumbled from the bursting of the stock price bubble in 2019.
- The Canadian market will continue to grow quickly.
- There are good hemp or cannabis business opportunities in Illinois, in red states with kind regulations, and internationally.

Exhibit hall at MJBizCon

The font size on the exhibit map for booth numbers was so small that I had to use my phone camera to enlarge the image for searching the map. The vendors in the exhibit hall represented the cannabis industry from seed to sale. For growers, there was lighting, nutrients, mildew prevention, water treatment, grow structures, potting material, and facility architects. The service industries of accountants, attorneys, and marketing were represented. Missing from the exhibit hall were laboratories, laboratory equipment, and transportation. Hardy Diagnostics was there with laboratory supplies. SGS was there with consulting services, and A2LA was there with laboratory accreditation. Attendees learned about safes and ATMs. There were numerous vendors for packaging of all cannabis products and equipment for filling and sealing of packaging. Attendees interested in extraction saw full-scale equipment for extraction and distillation. Micro- and nano-encapsulation of the cannabis extract was included. For edible manufacturers, vendors included candy molds, flavors, one chocolate supplier, and one nut butter supplier. A handful of edible manufacturers exhibited. While MJBizCon was not suitable for me finding my niche audience of edible manufacturers, MJBizCon is a great conference to attend for anyone on the learning curve of the cannabis industry like I am.

With 21 hours of total exhibit hall time, it was impossible to check out every one of the 1300 exhibitors. On Wednesday and the first day of the exhibit hall, I thoroughly covered four rows of exhibitors from a total of about 42 rows. Clearly the math was not in my favor, and simultaneous presentations competed for my time. In the afternoon of the second day of the exhibit hall, I created a list of 24 exhibitors I wanted to visit and saw half of those. I didn't think I would see the remaining half, but in the morning of the third and last day of the exhibit hall, I buzzed around the exhibit hall with my list before leaving the convention center at noon. The battle was won in the exhibit hall but the battle of listening to presentations was lost.

White label edibles

In speaking with a few edible vendors in the exhibit hall, I repeatedly heard the terms "white label" products and "co-packer." I pieced together that white label products are made by co-packers. I am not sure as to why I did not already know the definition of white label. The co-packer is a third-party facility that uses shared equipment to manufacture and package similar products from multiple companies and applies the label unique to each company. Co-packers are common in the food industry. Across the board, cannabis-infused edibles with Δ^9-tetrahydrocannabinol (THC) cannot be made in a food facility with equipment for making nonedibles. One vendor at MJBizCon told me that his CBD edible was made by a food co-packer and not by a stand-alone or stand-apart facility; I question the validity of that practice. It does make sense for cannabis start-up companies to use a co-packer until the company grows into their own facility. Before MJBizCon, I had not considered the use of co-packers in the cannabis industry, despite co-packers being of familiar use in the food industry.

Networking at MJBizCon

One of the best parts of attending a conference is networking in the exhibit hall, at conference sponsored events, in transit, and in the evening. People in the cannabis industry are passionate and love talking about their work and giving their opinion. I love hearing the stories of industry professionals as they share details and challenges. One story I heard involved international trade and the need to ship cannabis to Canada to meet the demand for cannabis in Canada. To test exporting to Canada, cannabis grown in Jamaica was being shipped to Canada in a small amount of 25 kg. The purpose of the shipment was to identify barriers for shipping into Canada. If successful, additional shipments of larger amounts will follow. Three stories came from an exuberant Uber driver. He detailed his personal experience with too much THC during a Seattle trip, going to church to get his magic brownies made with cannabis, and his in-home attempt at making magic brownies by directly adding the cannabis flower to the batter, which resulted in what he called "cannabis farts." You can't make this stuff up. Many of the people I spoke with had businesses that could have a booth in the exhibit hall, but the people chose to walk the exhibit hall, attend presentations, and network.

National Cannabis Industry Association Education Committee

At MJBizCon, I was able to meet in person the Chair of the National Cannabis Industry Association (NCIA) Education Committee. One of my greatest professional triumphs is being named to the NCIA Education Committee on behalf of ImEpik, LLC. ImEpik became a member of the NCIA in 2019 as I worked on revising ImEpik's online curriculum for Preventive Controls Qualified Individual. When NCIA called for volunteers to their committees, at first, I was not chosen. That was fine because the ImEpik CEO had been chosen for the Cannabis Manufacturing Committee. In September, I was contacted by NCIA asking if I would be on the newly created Education Committee. I was thrilled! The inaugural conference call of the Education Committee was in October 2019, and monthly conference calls continue.

The work of the committee is still forming, but the committee reviews abstracts for four NCIA conferences, identifies curriculum for training for cannabis jobs at colleges, communicates to the public about cannabis, and writes white papers on topics of concern to the cannabis industry. The committee is composed of industry leaders, and I am eager to work among such experts. I am most excited about developing curriculum for colleges to use in the training of adults for jobs in the cannabis industry, as this was my role at Northeast Wisconsin Technical College in Green Bay, Wisconsin, for 7 years.

In January 2020 I traveled to Denver, Colorado, with two of my fellow NCIA Education Committee members to the Community College of Denver (CCD) where we met with the President, Provost and Vice President for Academic Affairs, Dean of the Center for Health and Natural Sciences, and about a dozen, local cannabis industry members. NCIA is supporting CCD in the development of certificate, associate degree and eventually baccalaureate degree programs in Cannabis Science. Curriculum will be written, instructors will be hired, and students will be prepared to work in a sector of the cannabis industry along the seed-to-sale continuum. Everyone involved is listening to the needs of the cannabis industry and working to meet those needs. Across the nation, as courses are offered in cannabis at universities and colleges, best practices will be captured to train workers at the highest level. There is a great burden for on-the-job training in the cannabis industry, and it is good to know that educational institutions are working to bring training to the cannabis industry. I encourage industry membership in NCIA to benefit from this relationship with CCD among many other resources available from NCIA.

After-parties and late start

Through members of the NCIA Education Committee, I was invited to two evening parties. A colleague and I attended the Edibles Magazine party at the Palms Casino Resort. The suite on the 31st floor was sweet! It had a pool that jutted out from the building and huge windows for taking in the view of the Las Vegas city lights. The party included a full bar, box of rolled joints, great food (Thank you Jesus! We needed food.), dessert bar, DJ and dance floor, and a gift bag filled with goodies! Being a bit square and wanting to get up for the conference in the morning, we did not stay until the 2 a.m. party end time.

The second and third days of MJBizCon started at 9:30 a.m. The conference organizers know their audience and that attendance would be low with an 8 a.m. start time which is common in the food industry. Conference after-parties provided a venue for enjoying cannabis, the star of the show, given the strict adherence to the conference policy of no products with THC at the show (Fig. 13.2). Vendors with growing equipment brought poinsettias for their displays rather than cannabis because cannabis plants were not allowed. The availability of after-parties reminded me of the parties sponsored by food companies when I started in the food industry in the 1980s. The party atmosphere fits with the culture of cannabis.

I was a bit surprised that no complimentary food was offered by the conference. On Wednesday, I ordered at the single Starbucks store on-site at the convention center and then realized there was coffee available at a limited number of sites in the

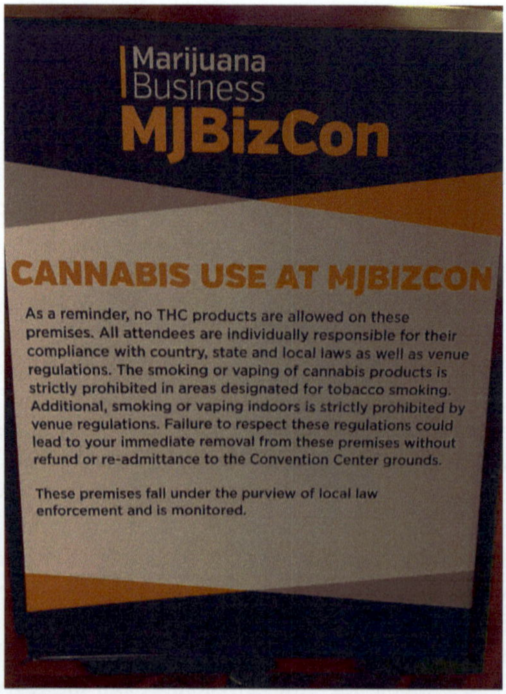

FIGURE 13.2

Cannabis use at MJBizCon.

Personal photo of the author.

morning, but no breakfast food. For lunch, there were many venues to purchase food within the exhibit hall and plenty of seating. I appreciated that there was plenty of space in which to sit in both the exhibit hall and the presentation rooms. During the reception in the exhibit hall from 4 to 6 p.m., wine and beer were available, but no food. I understand the logistics of serving 33,000 people is complicated. Not being a food or edible conference, there were very few snacks given away at the booths. This makes me appreciate the food industry conferences more in which booths have food samples and the conferences provide food. Food scientists do like food.

What did I learn about edibles at MJBizCon?

- The cannabis industry needs a conference dedicated to edibles. MJBizCon is a *business* conference. Edible manufacturers can source supplies and non-cannabis ingredients from the food industry and attend food industry conferences. Maybe they already do.
- Of the 1300 exhibitors, only a handful were edible manufacturers. About a dozen suppliers of flavors, other ingredients, equipment, or supplies to edible manufacturers exhibited.
- The cannabis-infused edibles industry is slowly scaling up from cannabis kitchens to pilot-plant-sized equipment as used in the food industry. In general, production

volume is not there to support the capital expense of automated equipment and then the double shift of production as is common in the food industry.

- Planet 13 dispensary in Las Vegas produces its own line of chocolate bars in a pilot-plant-sized processing facility on the same site as the dispensary.
- Cannabis-infused edible manufacturers are using co-packers.
- When the vape crisis hit in 2019, sales of vape cartridges decreased and sales of edibles increased.
- The demand from consumers is shifting and unpredictable among the cannabis products for smoke, vape, topicals, or edibles.
- Colorado is home to bigger companies in edibles.
- Do not purchase reserves of packaging because regulations will change and the company will be stuck with noncompliant packaging.
- Cannabiz Media maintains a current database of every licensed and active cannabis company in the United States.

Nevada field trip

I added another field trip to my list during my Nevada trip to MJBizCon. With dispensaries as common as Walgreens on every corner, it was not difficult to find a dispensary. My Uber driver recommended Planet 13 dispensary in Las Vegas. A colleague who has no personal experience with cannabis and I went to Planet 13 after dinner on our first day in Las Vegas. For me, it was a typical experience of registering with my driver's license, walking past armed security guards, and looking into well-lit display cases of products. Customers cannot touch cannabis products before purchasing. The room was spacious and held at least 14 display cases. I am guessing there were at least 50 people inside Planet 13, and it was not crowded. Planet 13 is certainly the brand new, shiny Cadillac of dispensaries and lived up to the recommendation. We were matched to an excellent budtender who walked the floor with us and patiently answered all our questions about cannabis in general and the cannabis products. Planet 13 sold flower, cartridges, shatter, topicals, and edibles.

I purchased Planet 13's homemade chocolate bar and learned it was made just outside the dispensary and down the hall in a spacious edible facility that I would not call a kitchen. It was the size of a food facility's pilot plant. From the hall, we looked through large glass windows at stainless steel equipment like two kids at the zoo and ogled over the plastic chocolate molds that fed into a filler. I appreciate that Planet 13 built in the observation of the production facility. That was genius!

Back in my hotel room, my colleague enjoyed one-half square of chocolate with a dose of 5 mg of THC and the next day reported no effect. I consumed one square with a dose of 10 mg of THC the first night and two squares the next night. For me, the effect takes about an hour and a half, so both nights I was nestled into bed before feeling any real effect. The second night I did feel a little tipsy when I woke up during the night. I will continue my field trips to dispensaries in 2020.

December 2019

This book is current as of December 2019 of my personal knowledge of the cannabis industry. Keeping up with cannabis regulations and cannabis information in general has been a game of Whack a Mole or the game at the fair where the targets are constantly moving as you try to knock just one down. Even if one target is hit, there are many others that go by. The information domestically is hard to keep up with, and information is pouring in internationally from all parts of the world. Cannabis is truly a global topic. Through the months of writing, I simply could not keep up the daily news on cannabis. In addition, the political climate in the United States certainly was distracting, with both impeachment hearings and the race for the democratic presidential candidate. I pushed myself to write quickly because I needed to capture information to a date and then just stop and walk away. As for my own education of cannabis, there have been times in which I thought I understood state code or best practices in the cannabis industry only to learn new information that contradicted my understanding or changed my perception. This tells me that practices support the label of the "wild, wild west" applied to the cannabis industry, meaning the cannabis industry is underregulated, is undersupervised, and includes rogue characters.

The popularity of vaping among youth was addressed head-on by FDA as a crisis in smoking, and 2019 saw a vape crisis traced to the addition of vitamin E oil to both nicotine and cannabis cartridges, which caused lung damage and deaths when smoked. Unfortunately, there will always be rogue characters in industry, trying to get by with substandard product, and regulation and enforcement may or may not catch the adulteration. I choose to believe that regulation and enforcement will help level the playing field among cannabis companies competing to sell a safe, quality product.

Politics

Seven democratic candidates remain on the debate stage in December 2019. The year 2020 could be a big year for the cannabis industry with a vote for legalization being on many state ballots and the potential to create policy and regulation at the federal level. In November 2019, it was reported in the press that Vice President Biden believes marijuana is a gateway drug to harder drugs. His campaign tried to walk that back with a reasonable statement calling for the need for further research. With few exceptions, it is generally agreed that cannabis is not addictive, unlike many prescription drugs prescribed legally. Everyone agrees that further research is needed, but the other presidential candidates have clearly stated that marijuana should be legalized, so Biden may take a hit there with the procannabis folks.

Legalization of marijuana

Legalization of marijuana at the federal level means that marijuana will be removed from the Drug Enforcement Administration Schedule I list of drugs. There is some talk of moving marijuana from Schedule I to Schedule II and regulating marijuana as a Schedule II drug and as a first step in legalization. It is only after descheduling that medical research with federal funds or within a federally funded facility will be universally allowed in the United States. It is only after descheduling of marijuana that the FDA will regulate cannabis-infused edibles with a THC content over 0.3%.

I have so many questions about the FDA regulation of cannabis-infused edibles.

- Will cannabis-infused edibles be regulated with existing FDA regulation for food and dietary supplements?
- Will new FDA regulation be written for edibles?
- Will FDA allow non-ready-to-eat (non-RTE) edibles that require a kill step by the consumer?
- Will FDA require environmental monitoring of the post-kill step area for RTE edibles?
- Will FDA allow refrigerated and frozen edibles of high water activity?
- Will FDA allow canned edibles?
- Will FDA allow edibles to be processed, manufactured, packaged, or held in a nonedible facility?
- Will FDA allow food co-packers to manufacture and package CBD edibles?
- Will FDA allow edibles to be transported with nonedibles?
- Will the FDA or United States Department of Agriculture (USDA) regulate meat edibles?
- Will the vape crisis push FDA to regulate?

The cannabis industry is doing an excellent job of preparing for and getting ahead of federal regulation. It makes sense that what is required by the FDA regulation now of medical devices, pharmaceuticals, foods, and dietary supplements must be implemented in the federally unregulated cannabis industry. I encourage cannabis-infused edible companies to stay ahead of future FDA regulations by implementing current FDA requirements and recommendations for food and dietary supplements. Then when FDA regulations are written, a cannabis company will just have to make minor changes to their programs for FDA compliance.

Food safety is not competitive

The food industry is highly competitive for the consumer's dollar, and many companies have a small margin of profit on the sale of their ingredient or food. However, competition among food companies does not enter the arena of food safety. We are all in food safety together to protect the consumer from illness or injury, and we all need each other to identify and share best practices. This must be true in the cannabis industry as well for the success of the cannabis industry. If one fails in the area of cannabis safety, it hurts the entire industry. Cannabis product and edible safety is not competitive. I encourage cannabis-infused edibles manufacturers to join their local affiliate of the International Association of Food Protection to network with local food safety experts in the food industry.

Conclusion

In my two years in the cannabis industry and preaching the good news of food safety, I have grown from the teenager who thinks she knows everything and naively believes she has all the answers on food safety applied to cannabis to the wise woman who realizes she has so very much to learn about the cannabis world. Food scientists like me are not required to understand the continuum of farm to fork, i.e., agriculture to foodservice; each food scientist excels in his/her expertise. As a cannabis scientist, I do not have that luxury and I strive to understand every sector of the business from seed to sale. The cannabis industry is composed of cannabis companies employing people who are working too hard and wearing too many hats with not enough time to develop expertise. Companies within states that require vertical integration are hardest hit, as the company is required by law to manage every step from seed to sale. Over time and without vertical integration, cannabis scientists will emerge with expertise in their business sector (Fig. 13.3).

Learning about cannabis is addictive. My personal education of the world of cannabis continues to be exhilarating. I will continue to read, network, publish, deliver training, conduct gap assessments, and investigate risk at cannabis facilities. With each person I meet and each facility I visit, I am better prepared for the next gig. As I continue to learn about cannabis, I am optimistic that the future is bright for cannabis-infused edibles (Box 13.1). As an adult living daily with an autoimmune disorder, I want to see every sector of the cannabis industry succeed and potentially offer an edible to alieve my symptoms.

FIGURE 13.3

Applying the cannabis industry practice of trimming and hanging to dry on wire hangers for the annual basil harvest.

Personal photo of the author.

Box 13.1 Real people in cannabis

Kathy Knutson, Consultant, independent contractor, lead instructor, project manager, food scientist, microbiologist, author ... cannabis scientist.

Time in the cannabis industry: 2 years.

What drives me in my job: In 2017 as a food scientist, I was looking to build my career and learned of work in the cannabis industry. I am now addicted to learning about the cannabis industry by working with clients in the cannabis industry, publishing, speaking, attending conferences, and networking. I expect to be very busy with work in the cannabis industry and want to be fully prepared to help my clients implement best practices toward producing a product that is safe for any consumer, but most importantly, the most vulnerable immunocompromised consumer.

Reference

[1] Chart: Cannabis industry, MJBizCon continue growth trends. https://mjbizdaily.com/chart-cannabis-industry-mjbizcon-continue-growth-trends/. [Accessed 12.19.19].

Appendix 1

Outline of Good Manufacturing Practices (GMPs) for the cannabis industry. The outline was garnered from the alignment of food and dietary supplement GMPs and lists programs, procedures, and policies that provide for a product that is safe. Management is responsible for the design of programs, policies, and procedures; writing of Standard Operating Procedures and records; and training. Without federal oversight, the cannabis industry is not provided with GMPs from FDA and must look to current FDA resources for GMPs. The outline is divided into responsibilities of management and responsibilities of employees at cannabis facilities. Details exist for each topic, and additional resources should be consulted.

A. Absolutes of GMPs at any cannabis facility. Management is responsible for the design of programs, policies, and procedures.
  a. Site security
    i. Employee entrance
    ii. Visitor entrance
    iii. Cameras
    iv. Barriers
    v. Receiving
    vi. Shipping
    vii. Storage of ingredients
    viii. Holding of product
    ix. Transportation
      1. Manifest
  b. Ethical enforcement of state code
    i. Employees may not be compensated with product
    ii. Employees may not be incentivized with product
    iii. It is illegal for employees to take product
  c. Employee training
    i. On-boarding training
    ii. Routine refresher training
    iii. Training for cause or to a new task
    iv. Integrity and professionalism
    v. The immunocompromised patient or consumer

vi. The importance of records and recordkeeping
vii. Biological hazards
    1. *Salmonella*
    2. Pathogenic *Escherichia coli*
    3. *Listeria monocytogenes*
viii. Chemical hazards
    1. Pesticides
    2. Heavy metals
    3. Residual solvents
    4. Mycotoxins
    5. Allergens
ix. Physical hazards
x. Disease control
xi. Employee hygiene and cleanliness policies
xii. Prevention of cross-contact and cross-contamination
    1. Handwashing
    2. Personal protective equipment
    3. Traffic patterns
xiii. Allergen management
xiv. Standard operating procedures
xv. Basics of cleaning and sanitizing
    1. Sanitation Standard Operating Procedures

d. Allergen management program for edibles facilities
e. Disease control
    i. Supervisory observation of employees working with ingredients, product, food contact surfaces, or product packaging materials
    ii. Policy and procedure for reporting illness, open lesion, including boils, sores, or infected wounds, or any abnormal source of microbial contamination
f. Employee hygiene and cleanliness policies
    i. Personal hygiene
        1. Oral hygiene
        2. Control of perspiration
        3. Cosmetics
        4. Topicals
        5. Fingernails
    ii. Jewelry
    iii. Storage of personal belongings
    iv. Handwashing
    v. Personal protective equipment
        1. Sanitary and intact gloves
        2. Sanitary shoes or boots
        3. Hair restraints including body, beard, and head

g. Policies and areas for employees
   i. Storing of personal belongings
   ii. Toileting
   iii. Eating food
   iv. Drinking beverages
   v. Chewing gum
   vi. Using tobacco
h. Maintenance of grounds
   i. Solid waste disposal
      1. Dumpsters
   ii. Landscaping
      1. Removal of harborage sites for pests
      2. Removal of litter
   iii. Adequate drainage
   iv. Roads and parking lots
   v. Sanitary sewer and sewage treatment and disposal
   vi. Security barriers
i. Sanitary facility design
   i. Preventive maintenance schedule
   ii. Adequate space
   iii. Prevention of contamination
      1. Separation of operations
      2. Prevention of cross-contact at edibles facilities
   iv. Protection of outdoor bulk vessels
   v. Adequate cleaning
   vi. Adequate lighting
   vii. Adequate ventilation
      1. Odor abatement
   viii. Adequate screening against pests
j. Sanitary operations
   i. Storage of chemicals for cleaning and sanitizing
   ii. Pest control management
   iii. Sanitation of product contact surfaces
   iv. Sanitation of nonproduct contact surfaces
   v. Storage of cleaned and sanitized portable equipment and utensils
k. Sanitary facilities and controls
   i. Sanitarian
   ii. Sanitation master schedule
   iii. Solid waste removal from facility
      1. Traffic patterns
   iv. Water potability
      1. Source of ice

- v. Adequate and maintained plumbing
- vi. Sewage disposal
- vii. Toilet facilities
- viii. Handwash stations
- ix. Gowning station
- x. Receiving at facilities
  1. Prevent water, waste, and uncontrolled air ingress
  2. Spill cleanup
- xi. Storage
  1. Spill cleanup
  2. Raw plant material
  3. Chemicals, nonproduct supplies, packaging, and labels
  4. Ingredients
     a. Protective containers
  5. Rework
     a. Label
  6. Freezers and Coolers
     a. Temperature recording
     b. Humidity control
  7. Product
     a. Onsite
     b. Warehousing
- xii. Transportation vehicles
  1. Receiving ingredients
  2. Distribution of products

l. Equipment and utensils
- i. Preventive maintenance schedules
- ii. Hygienically designed holding and conveyance systems
- iii. Hygienically designed and installed equipment
  1. Smooth seams
- iv. Intact tools and utensils of hygienic design
- v. Compressed air and gases

m. Establish a production and process control system
- i. Quality control
  1. Instruments and controls
     a. Calibration
     b. Accuracy checks
     c. Maintenance
- ii. Receiving of components (ingredients) and packaging and labeling for products
  1. Safety of packaging as a product contact surface

- iii. Master Manufacturing Record and Batch Production Record
  - 1. Extraction facilities
  - 2. Edibles facilities
- iv. Laboratory operations
  - 1. Chemical analysis
    - a. Pesticides
    - b. Heavy metals
    - c. Residual solvents
    - d. Mycotoxins
  - 2. Microbial analysis
    - a. Aerobic Plate Count
    - b. Total Yeast and Mold Count
    - c. Pathogens
  - 3. Extraneous material analysis
  - 4. Internal laboratory
  - 5. Third-party laboratory
- v. Manufacturing operations
  - 1. Extraction facilities
  - 2. Edibles facilities
    - a. Prevention of pathogen growth during manufacturing
- vi. Packaging and labeling operations

n. Returned product
o. Product complaints
p. Records and Recordkeeping
- i. Requirements applying to records

B. Absolutes of employee GMPs at any cannabis facility. Employees are responsible for following programs, policies, and procedures.
a. Reporting of illness, open lesion, including boils, sores, or infected wounds, or any abnormal source of microbial contamination
b. Employee hygiene and cleanliness policies
- i. Personal hygiene
  - 1. Oral hygiene
  - 2. Control of perspiration
  - 3. Cosmetics
  - 4. Topicals
  - 5. Fingernails
- ii. Jewelry
- iii. Storage of personal belongings
- iv. Handwashing
- v. Personal protective equipment
  - 1. Sanitary and intact gloves
  - 2. Sanitary shoes or boots
  - 3. Hair restraints including body, beard, and head

- c. Cybersecurity and data protection
  - i. Process controls
  - ii. Patient databases
  - iii. Company records
- d. Prevention of cross-contamination and cross-contact
  - i. Site security
  - ii. Restricted access
    - 1. Key cards or biometrics
  - iii. Foot and wheeled equipment traffic patterns
- e. Recordkeeping
- f. Packaging and labeling

C. Absolutes of employee GMPs at a cannabis growing operation. Employees are responsible for following programs, policies, and procedures.
- a. Sanitary facilities and controls unique to growing
  - i. Field growing
  - ii. Outdoor greenhouse
  - iii. Indoor greenhouse
- b. Receiving of supplies
- c. Harvesting
  - i. Trim waste management and disposal
- d. Drying
  - i. Trim waste management and disposal
  - ii. Sanitary facilities and controls unique to drying

D. Absolutes of employee GMPs at a cannabis extraction facility. Employees are responsible for following programs, policies, and procedures.
- a. Sanitary facilities and controls unique to extraction
- b. Receiving
  - i. Receiving of raw plant material
- c. Master Manufacturing Record
- d. Batch Production Record

E. Absolutes of employee GMPs at a cannabis-infused edibles facility. Employees are responsible for following programs, policies, and procedures.
- a. Sanitary facilities and controls unique to edibles
- b. Procedures for control of hazards and to prevent cross-contact, cross-contamination, and growth
  - i. Measures to destroy or prevent the growth of pathogens
  - ii. Processing steps
  - iii. Rework
  - iv. Work in process: holding of batters, breading, sauces, gravies, dressings, or dipping solutions
  - v. Control of water
    - 1. Monitoring of water activity

- vi. Acid and acidified food
  - 1. Monitoring of acidity
    - a. pH 4.6
- c. Master Manufacturing Record
- d. Batch Production Record
- e. Laboratory operations
  - i. Reserve samples and retains
- f. Recordkeeping
- g. Allergen management program
  - i. Eating food
    - 1. Vending
    - 2. Personal food and beverages
  - ii. Fresh PPE for each edible without unique allergens

Appendix 2

Product Name: Cannabis-infused Dark Chocolate Coconut Almond Toffee Bar

Product Description: Dark chocolate is melted, and cannabis extract is added. Toffee is prepared with toasted coconut, toasted almonds, butter, corn syrup, and sugar. After the chocolate has cooled in molds, cooled toffee is added as a topping. The water activity of the bar is below 0.6, making the bar shelf-stable.

Hazard identification (column 2) considers those that may be present in the food; because the hazard occurs naturally, the hazard may be unintentionally introduced, or the hazard may be intentionally introduced for economic gain.

B = Biological hazards including bacteria, viruses, parasites, and environmental pathogens

C = Chemical (including radiological) hazards, food allergens, substances such as pesticides and drug residues, natural toxins, decomposition, and unapproved food or color additives

P = Physical hazards include potentially harmful extraneous matter that may cause choking, injury, or other adverse health effects

Hazard Analysis

(1) Ingredient/ Processing Test		(2) Identify potential food safety hazards introduced, controlled, or enhanced at this step	(3) Do any potential food safety hazards require a preventive control? Yes	No	(4) Justify your decision for column 3	(5) What preventive control measure(s) can be applied to significantly minimize or prevent the food safety hazard? Process, Allergen, Sanitation, Supply chain, or Other preventive control	(6) Is the preventive control applied at this step? Yes	No
Water Source	B	Waterborne pathogens		X	The hazard is controlled by GMPs.			
	C	Heavy metals, industrial chemicals, radioactive chemicals, pesticides		X	The hazard is controlled by GMPs.			
	P	Rusted pieces of pipes		X	The hazard is controlled by GMPs.			
Receiving Packaging	B	None						
	C	Undeclared allergens	X		Product label must identify allergens.	Allergen preventive control	X	
	P							
Receiving Refrigerated Ingredients Butter (Milk Fat)	B	Growth of pathogens and/or toxin formation		X	The hazard is controlled by GMPs at the supplier.			
		Vegetative pathogen survival of pasteurization	X		Pathogens are associated with raw milk.	Process preventive control		X
		Environmental pathogens	X		Post-pasteurization steps must control for *Listeria monocytogenes*.	Process preventive control		X
	C	Drug residues		X	Farmers use antibiotics to control cow illness. Milk is tested for antibiotics before off-loading into silo at the supplier.			

		Undeclared allergen	X		Allergen cross-contact must be prevented. The butter factory does make flavored butter with additional allergens.	Supply Chain preventive control	X	
			X		Incorrect label on butter.	Supply Chain preventive control	X	
			X		The finished product label must declare the milk allergen.	Allergen preventive control		X
	P	Metal inclusion	X		The butter making process includes metal-on-metal.	Supply chain preventive control	X	
Receiving Ambient Ingredients Coconut, shredded	B	Vegetative pathogens	X		Coconut is a raw agricultural commodity.	Process preventive control		X
		Environmental pathogens	X		Post-kill steps must control for pathogens.	Supply chain preventive control	X	
	C	Undeclared allergen	X		The finished product label must declare the tree nut allergen.	Allergen preventive control		X
	P	Husk pieces		X	The hazard is controlled in shredding at the supplier reducing the likelihood.			
		Metal inclusion	X		The coconut shredding process includes metal-on-metal, and the supplier has a history of metal in the ingredient.	Supply chain preventive control	X	
Receiving Ambient Ingredients Corn Syrup	B	None						
	C	Unapproved colors and additives		X	The hazard is controlled by GMPs at the supplier.			
	P	None						
Receiving Ambient Ingredients Sugar, dry	B	None						
	C	Unapproved colors and additives		X	The hazard is controlled by GMPs at the supplier.			
	P	None						

Continued

—cont'd

Hazard Analysis

(1) Ingredient/ Processing Test	(2) Identify potential food safety hazards introduced, controlled, or enhanced at this step		(3) Do any potential food safety hazards require a preventive control?		(4) Justify your decision for column 3	(5) What preventive control measure(s) can be applied to significantly minimize or prevent the food safety hazard? Process, Allergen, Sanitation, Supply chain, or Other preventive control	(6) Is the preventive control applied at this step?	
			Yes	No			Yes	No
Receiving Ambient Ingredients Dark Chocolate, solid	B	Vegetative pathogen survival of kill step(s)	X		Pathogens are associated with raw cocoa.	Supply chain preventive control	X	
		Environmental pathogens	X		Post-kill steps must control for *Salmonella*.	Supply chain preventive control	X	
	C	Heavy metal		X	The supplier has no history of the hazard of heavy metals in their ingredients.			
		Unapproved colors and additives		X	The hazard is controlled by GMPs at the supplier.			
		Undeclared allergen	X		Allergen cross-contact must be prevented. The chocolate factory does make chocolate with additional allergens, e.g. tree nuts. All chocolate is considered to contain the milk allergen.	Supply chain preventive control	X	
			X		Incorrect label	Supply chain preventive control	X	
			X		The finished product label must declare the milk allergen.	Allergen preventive control		X
	P	Metal inclusion	X		The chocolate and bar making process includes metal-on-metal.	Process preventive control		X

Receiving Ambient Ingredients Almonds, roasted and whole	B	Vegetative pathogens	X		Pathogens are associated with raw nuts. The hazard will be controlled at the Cook Topping step.	Process preventive control		X
		Environmental pathogens	X		Post-roasting steps must control for *Salmonella*.	Supply chain preventive control	X	
	C	Undeclared allergen	X		Allergen cross-contact must be prevented. The nut factory does roast other tree nuts.	Supply chain preventive control	X	
			X		Incorrect label.	Supply chain preventive control	X	
			X		The finished product label must declare the tree nut allergen.	Allergen chain preventive control		X
	P	Metal inclusion	X		The roasting process includes metal-on-metal. (The nut factory does not pack into glass.)	Supply chain preventive control	X	
Receiving Ambient Ingredients Cannabis Extract	B	Pathogenic *Escherichia coli* such as Shiga-toxin producing *E. coli* (STEC)		X	The decarboxylation process is validated to kill vegetative pathogens, and the extract is released per state code.			
		Salmonella		X	The decarboxylation process is validated to kill vegetative pathogens, and the extract is released per state code.			
		Total Yeast and Mold		X	TYMC is an indicator test only and does not control for a hazard.			
		Environmental pathogens	X		Post-extraction steps must control for pathogenic *E. coli, Salmonella* and *Listeria monocytogenes*.	Supply chain preventive control	X	
	C	Pesticides	X		Cannabis has a history of banned pesticides or high pesticide residues.	Supply chain preventive control	X	

Continued

—cont'd

Hazard Analysis

(1) Ingredient/ Processing Test	(2) Identify potential food safety hazards introduced, controlled, or enhanced at this step	(3) Do any potential food safety hazards require a preventive control? Yes	No	(4) Justify your decision for column 3	(5) What preventive control measure(s) can be applied to significantly minimize or prevent the food safety hazard? Process, Allergen, Sanitation, Supply chain, or Other preventive control	(6) Is the preventive control applied at this step? Yes	No
				Pesticides must be tested per state code.			
	Mycotoxins	X		Cannabis has a history of high mold counts in flower and trim. If the cannabis extract is sourced from retail marijuana flower or trim that failed microbial testing, mycotoxins must be tested per state code.	Supply chain preventive control	X	
	Clostridium botulinum toxin		X	*Clostridium* spores are not known to germinate and grow in a low water activity ingredient such as extract. The extract is refrigerated and temperature is monitored as a GMP.			
	Nonfood grade extraction solvent and residual solvents	X		Residual solvents must be tested per state code.	Supply chain preventive control	X	
	Heavy metal	X		Cannabis has a history of high levels of heavy metals based on the location of growth. Metals must be tested per state code.	Supply chain preventive control	X	

Receiving Ambient Ingredients Cannabis Extract (continued)	C	Other contaminants per state code	X		Test for other contaminants per state code.	Supply chain preventive control	X	
	P	None						
Storage Refrigerated Ingredients	B	None			Temperature is monitored as a GMP.			
	C	None						
	P	None						
Storage Ambient Ingredients	B	None						
	C	None						
	P	None						
Secure Storage Cannabis Extract	B	None			Temperature is monitored as a GMP.			
	C	None						
	P	None						
Rework	B	None			Only chocolate that has remained at and controlled within the facility can go back into Melt.			
	C	Undeclared allergen		X	The allergen management program will control allergens from rework.			
	P	None						
Melt	B	Environmental pathogens	X		*Salmonella* and *Listeria monocytogenes* must be controlled in the processing environment and on equipment surfaces.	Sanitation preventive control	X	
		Pathogen growth during Melt		X	Temperature is monitored as a GMP.			
	C	None			No unique allergens.			
	P	Metal inclusion	X		The Melt process includes metal-on-metal.	Process preventive control		X

Continued

—cont'd

Hazard Analysis

(1) Ingredient/ Processing Test	(2) Identify potential food safety hazards introduced, controlled, or enhanced at this step		(3) Do any potential food safety hazards require a preventive control?		(4) Justify your decision for column 3	(5) What preventive control measure(s) can be applied to significantly minimize or prevent the food safety hazard? Process, Allergen, Sanitation, Supply chain, or Other preventive control	(6) Is the preventive control applied at this step?	
			Yes	No			Yes	No
Screen and Magnet	B	Environmental pathogens	X		*Salmonella* and *Listeria monocytogenes* must be controlled in the processing environment and on equipment surfaces.	Sanitatiation preventive control	X	
	C	None			No unique allergens.			
	P	Metal Inclusion	X		Previous process steps include metal-on-metal.	Process preventive control	X	
Temper	B	Environmental pathogens	X		*Salmonella* and *Listeria monocytogenes* must be controlled in the processing environment and on equipment surfaces.	Sanitation preventive control	X	
	C	None			No unique allegens.			
	P	Metal Inclusion	X		Temper includes metal-on-metal.	Process preventive control		X
Mix	B	Environmental pathogens	X		*Salmonella* and *Listeria monocytogenes* must be controlled in the processing environment and on equipment surfaces.	Sanitation preventive control	X	
	C	None						
	P	Metal inclusion	X		Mix includes metal-on-metal.	Process preventive control		X

Toast Almonds	B	Vegetative pathogens		X	This is not a validated kill step.		
	C	Undeclared allergen	X		Allergen cross-contact must be prevented.	Sanitation preventive control	X
	P	None					
Toast Coconut	B	Vegetative pathogens		X	This is not a validated kill step.		
	C	Undeclared allergen	X		Allergen cross-contact must be prevented.	Sanitation preventive control	X
	P	None					
Cook Topping	B	Vegetative pathogens	X		Cooking to 290°F kills potential pathogens in the ingredients for the topping.	Process preventive control	X
	C	Undeclared allergen	X		Allergen cross-contact must be prevented.	Sanitation preventive control	X
	P	None					
Mold	B	Environmental pathogens	X		*Salmonella* and *Listeria monocytogenes* must be controlled in the processing environment and on equipment surfaces.	Sanitation preventive control	X
	C	Undeclared allergen	X		Allergen cross-contact must be prevented.	Sanitation preventive control	X
	P	None					
Cool Chocolate and Topping and Add Topping	B	Environmental pathogens	X		*Salmonella* and *Listeria monocytogenes* must be controlled in the processing environment and on equipment surfaces.	Sanitation preventive control	X
	C	None					
	P	None					
Demold	B	Environmental pathogens	X		*Salmonella* and *Listeria monocytogenes* must be controlled in the processing environment and on equipment surfaces.	Sanitation preventive control	X

Continued

—cont'd

Hazard Analysis

(1) Ingredient/ Processing Test		(2) Identify potential food safety hazards introduced, controlled, or enhanced at this step	(3) Do any potential food safety hazards require a preventive control? Yes	No	(4) Justify your decision for column 3	(5) What preventive control measure(s) can be applied to significantly minimize or prevent the food safety hazard? Process, Allergen, Sanitation, Supply chain, or Other preventive control	(6) Is the preventive control applied at this step? Yes	No
		Pathogens associated with human illness	X		Proper handwashing, personal protective equipment (PPE), and exclusion of ill workers are required.	Sanitation preventive control	X	
	C	None						
	P	None						
Inspect and Wrap	B	Environmental pathogens	X		*Salmonella and Listeria monocytogenes* must be controlled in the processing environment and on equipment surfaces.	Sanitation preventive control	X	
		Pathogens associated with human illness	X		Proper handwashing, PPE, and exclusion of ill workers are required.	Sanitation preventive control	X	
	C	Undeclared allergen	X		Allergen cross-contact must be prevented.	Sanitatiaon preventive control	X	
	P	None						
Metal Detector	B	None						
	C	None						
	P	Metal inclusion	X		Previous process steps include metal-on-metal.	Process preventive control	X	

Hold in Secure Storage ***temperature control***	B	None		Temperature is monitored as a GMP.		
	C	None				
	P	None				
Label, Box and Case	B	None				
	C	Undeclared allergens	X	The correct label must be applied to the product.	Allergen preventive control	X
	P	None				
Hold in Secure Storage ***temperature control***	B	None		Temperature is monitored as a GMP.		
	C	None				
	P	None				
Transport	B	None				
	C	None				
	P	None				

Index

'*Note*: Page numbers followed by "f" indicate figures, "t" indicate tables and "b" indicate boxes.'

M

N

O

R

S